ENCYCLOPAEDIC DICTIONARY OF SCIENCE

ENCYCLOPAEDIC DICTIONARY OF SCIENCE

Vol. 2

[D—K]

Edited by

SATISH ANAND

ANMOL PUBLICATIONS PVT. LTD.

NEW DELHI-110 002

ANMOL PUBLICATIONS PVT. LTD.
4374/4B, Ansari Road Daryaganj
New Delhi-110 002

Encyclopaedic Dictionary of Science

First Edition 1995
Reprint 1997
ISBN 81-7488-003-8(Set)

PRINTED IN INDIA

Published by J.L. Kumar for Anmol Publications Pvt. Ltd., New Delhi-110 002 and Printed at Mehra Offset Press, Delhi.

Preface

The Encyclopaedic Dictionary of Science is an effort to keep pace with continuing rapid developments and expanding vocabularies in various branches of science. This dictionary has been encyclopaedic in its contents and style of representation. Throughout the process of compilation and editing the volumes, the main effort was to make the dictionary serve as a ready reference for undergraduate and postgraduate students of science and for those appearing for competitive examinations, research scholars, teachers and authors. The entries have been written in a clear and explanatory style to provide both straight forward definitions and invaluable background information. This approach, combined with an extensive cross-reference system, enables the reader to place the each entry into a broader scientific context. At certain appropriate places, the diagrams have been included whenever the meaning of a word can be best understood by means of a diagram.

The presentation throughout has been aimed at sustaining the interest of readers while enriching their vocabulary and comprehension of technical terms and expressions of various branches of science.

This encyclopaedic dictionary will be of immense value to the students of science and to scientists and research scholars working on science projects.

In compiling a set of this kind it becomes necessary to draw upon the work of many authorities and seek the advice of colleagues to all of whom the editor is deeply indebted. All the comments from users on omissions or shortcomings will be most welcome.

The Editor

Vol. 2
D — K

Dacron. ® A type of polyester resin used in synthetic fibres.

Daddy longlegs or harvestman. An Arachnid with long slender legs, related to the Spiders. It has eight legs extending from a rounded or oval body. The daddy longlegs is omnivorous, feeding on plant fluids, animal tissue, and other daddy longlegs. The unrelated long-legged crane fly (an Insect) is also called daddy longlegs.

Dahlia. Tuberous-rooted, perennial plant (genus *Dahlia*) of the Composite family, native to Guatemala and Mexico, and widely cultivated. Most of the thousands of horticultural varieties are developed from one species (*D. pinnata),* the garden dahlia. Dahlias are stout, woody plants with colorful, late-blooming flowers. The tubers of the garden dahlia are a source of fructose used by diabetics.

Dairying. Industry concerned with producing, processing and distributing Milk and milk products. Ninety percent of the world's milk comes from Cattle; the rest comes from Goats, Buffalo, Sheep, Reindeer, Yaks, and other ruminants. About one third of the milk is used from Butter, an almost equal amount for market milk, and the rest for farm uses and the making of Cheese, concentrated milks, Ice Cream, and such by-products as dried milk solids (e.g., Lactose and Casein). The rise of modern, large-scale dairying paralleled the growth of urban populations and was stimulated by the invention of the cream separator and other specialized machines, the discovery of Pasteurization, and the study of cattle breeding.

Daisy. Name for several common wild flowers of the Composite family. The true, or English, daisy (*Bellis perennis*) is cultivated in the U.S.; its white, pink, or red flowers close at night. Other species include the purple Western daisy (*Astranthum or B. integrifolium*); and the common, white, or oxeye, daisy of the U.S. (*Chrysanthemum leucanthemum)*. Other plants, called daisy include the yellow daisy, or Black-eyed Susan; the Michaelmas daisy, an Aster; the Paris daisy, one of the plants also called marguerite; and the seaside daisy and daisy fleabane of the fleabane genus.

Dale, Sir Henry Hallett, 1875-1968. English scientist. For his study of Acetylcholine as agent in the chemical transmission of nerve impulses, he shared with Otto Loewi the 1936 Nobel Prize in physiology or medicine. Dale also investigated the pharmacology of ergot and histamine shock. He was knighted in 1932.

Dalmatian. Hardy Nonsporting Dog, known for speed and endurance; shoulder height, 19-23 in. (48.3-58.4 cm); weight, 35-50 lb (15.9-22.7 kg). It has a short, glossy coat, white with black of dark-brown spots. Probably developed in Dalmatia several hundred years ago, it was associated with horse and with horse-drawn coaches and fire engines; it remains a firemen's mascot.

Dalton. *See* atomic mass unit. [After John Dalton (1766-1844), British chemist.]

Dalton's atomic theory. The first modern attempt (1808) at an explanation of the behaviour of chemical elements. Dalton postulated that every element consists of indivisible particles called atoms. All atoms of a given element are identical and atoms can neither be created nor destroyed. Compound elements (compounds) are formed by combination of atoms of different elements in simple ratios to form compound atoms (molecules). The theory was able to interpret the laws of chemical combination and the law of conservation of mass.

Dalton, John. 1766-1844, English scientist. He taught mathematics and physical sciences at New College, Manchester. Dalton revived the atomic theory of matter, which he applied to a table of atomic weights and used in developing his law of partial pressures (Dalton's law). He was color-blind and studied that affiliction, also known as Daltonism.

Dalton's law of partial pressures. The principle that at constant temperature the total pressure exerted by a mixture of (ideal) gases in a definite volume is equal to the sum of the individual pressures each gas would exert if it occupied the same total volume alone: i.e. $p = p_1 \; p_2 + p_3$ .... where the individual pressures, p_1, etc., are the *partial pressures* of the gases.

Damping. The progressive decrease in amplitude of oscillations with time. An oscillating system is damped by forces caused by viscosity, friction, magnetic fields, etc. The resisting force is approximately proportional to the velocity. Three types of damping are distinguished according to the way in which the system returns to its equilibrium position when displaced. If it performs a large number of dying vibrations, it is said to be *underdamped.* An *overdamped* system, on the other hand returns to equilibrium slowly without oscillation. If the system just fails to oscillate it returns to equilibrium in its minimum time: it is then said to be *critically damped.*

Dandelion. Perennial herb (genus *Taraxacum*) of the Composite family, widely distributed in temperate regions. It has a rosette of deep-toothed leaves and a bright yellow head followed in fruit by a round head of white down for wind distribution. The common dandelion (*T. officinale*) is considered a lawn pest in the U.S. but is cultivated for food and medicine elsewhere. The leaves are used in salads and as a seasoning; the roots are used as a coffee substitute, as a bitter tonic, and as a laxative; the flower heads are used for making dandelion wine and as forage for bees. The dandelion plant contains a white latex—a hydrocarbon—that can be converted into a combustible fuel. One species is grown commercially for latex in the USSR and Argentina.

Daniell Cell. A type of primary cell in which the positive electrode is a copper rod immersed in copper sulphate solution and the negative electrode is an amalgamated zinc rod in dilute sulphuric acid, the two electrolytes being separated by a porous pot. At the positive electrode copper ions gain electrons from the rod and copper is deposited. At the negatide electrode zinc atoms dissolve as zinc ions. The e.m.f. is about 1.08 volts. [After J.F. Daniell (1790-1845), British physicist].

Daraf. A unit of elastance: the reciprocal of the farad.

Darwin, Charles Robert, 1809-82. English naturalist, grandson of Erasmus Darwin. He firmly established the theory of organic Evolution. His position as official naturalist aboard the H.M.S. Beagle during its world voyage (1831-36) started Darwin on a career of accumulating and assimilating data that resulted in the formulation of his concept of evolution. In 1858 he and Alfred Russel Wallace simultaneously published summaries of their independently conceived notions of Natural Selection; a year later Darwin set forth the structure of his theory and massive support for it in his *Origin of Species.* This was supplemented by later works, notably *The Descent of Man* (1871). Darwin also formulated a theory of the origin of coral reefs.

Dark-field Illumination. Illumination of a sample in a microscope by light from the side rather than underneath. Small specimens are then more easily seen as bright objects on a dark background.

Dasymeter. An instrument for determining the density of gases by measuring the upthrust produced on a tube of known volume.

Data Processing or **Information processing.** The operations of handling, merging, sorting, and computing performed upon data in accordance with strictly defined procedures, e.g., recording and summarizing the financial transactions of a business. In automatic data processing the operations are performed by a Computer. In distributed data processing some or all of the computer facilities are located in different places and connected by telecommunications links.

Date. Name for a Palm (*Phoenix dactylifera*) and for its edible fruit, probably native to Arabia and North Africa and long a major food source in desert and tropical regions. The sweet, nutritious fruits grow in heavy clusters; a tree may produce up to 200 lb (90 kg) annually. Sugar and a fermented drink are made from the sap of the tree, and the seeds are sometimes used as a coffee substitute. The wood of the trunk is used in construction, and mats and baskets are woven from the leaves.

Dating. In geology, archaeology, paleontology, and physics, the determination of the actual or relative age of an object, of a natural phenomenon, or of a series of events. The most common and accepted method of absolute geologic dating (establishment of actual age) is based on the natural Radio-Activity of certain minerals found in rocks; because the rate of radioactive decay of any particular Isotope is known, the age of a specimen can be computed from the ratio of the remaining isotope and its decay product. Relative geological dating (determination of the sequence of geological events) is accomplished by marking out the succession in which rocks were deposited, using various surface criteria. The identification of Fossils in a geological sample is very useful, because certain assemblages of species are characteristic of specific geologic time periods. Pollen analysis, or palynology, can also establish the succession of deposits as well as the climate prevalent at the time of deposition. In archaeology and recent geology, the carbon-14 method, which measures the remaining amount of this radioactive isotope in organic matter, can date specimens as old as 35,000 years. The death of an organism terminates the previous incorporation of this isotope (created by Cosmic-Ray bombardment) at a rate equal to the isotope's known decay rate. This method's accuracy has been improved by cross-checking with dendrochronology, in which the age of a piece of ancient wood can be determined by examining the ring pattern in its cross section. The pattern of individual ring widths in a sample reflects local climatic conditions occurring at the time when the piece of wood was a growing tree. By the use of overlapping patterns found in different wood samples, dating by this method has been carried back over 8,000 years.

Dative Bond. *See* Coordinate bond.

Daughter. A nuclide produced by decay of another nuclide (the *parent).*

Davisson-Germer experiment. An experiment in which a beam of electrons was directed onto the plane face of a single crystal of nickel in a high vacuum and the angular distribution of reflected electrons was investigated. It was found that the electrons were diffracted by the lattice and showed a maximum intensity at one particular angle. The experiment, performed in 1927, demonstrated the existence of the de Broglie waves predicted three years earlier.

Davy, Sir Humphry, 1778-1829. English chemist and physicist. He was professor (1802-13) at the Royal Inst., London, where his electrochemical researches led to his isolation of Sodium and Potassium (1807) and Barium, Boron, Calcium, and Magnesium (1808). He showed chlorine to be an element, theorized that acids characteristically contain hydrogen, and classed chemical affinity as an electric phenomenon.

Davy Lamp. A safety oil lamp used by miners. The flame burns inside a metal mesh cage or mantle, which cools the hot gases from the flame by conduction, preventing the flame from igniting any inflammable gases outside the lamp. The nature of the flame can indicate the presence of flammable gas. [Invented by Sir Humphrey Davy (1778-1829), English scientist.]

Day. The time taken for the earth to make one rotation on its axis. It is measured in various ways.

Daylight Saving Time (DST). Time observed when clocks and other timepieces are set ahead, usually by 1 hr. so that the sun will rise and set later in the day as measured by civil time. DST conserves lighting power and provides more usable daylight hours for afternoon and evening activities. First adopted during World War I by the U.S. and other countries, DST in the U.S. currently extends from the last Sunday in April to the last Sunday in October, the interval when daylight periods are longest.

DDT (Dichlorodiphenyltrichloroethane). A colourless crystalline solid, $(ClC_6H_4)_2\text{-}CHCCL_3$, with a slight aromatic odour, manufactured by the condensation of chloral with chlorobenzene. It is a powerful insecticide and has been widely used in agriculture, although this has now been restricted.

Deafness. Partial or total loss of hearing. It may be present at birth (congenital) or acquired any time thereafter. Conductive deafness, one of the two major types of deafness, involves a disturbance in the transmission of sound to the nerve receptors of the inner ear. This type of deafness may be caused by infection, impacted wax, perforation of the eardrum, or otosclerosis, a chronic condition, especially in older people, that restricts the vibration of the bone leading to the inner ear. These conditions can usually be treated with Antibiotics, surgery hearing aids, and other techniques. The other major type of deafness—perceptive, or nerve, deafness—involves damage to neural receptors in the inner ear, to nerve pathways to the brain, or to the area of the brain concerned with hearing. Usually permanent, it can be caused by infection, senility, tumors, or excessive noise, or it may be congenital. Those who cannot be helped to hear may communicate through lip reading and Sign Language. The first U.S. public school for the deaf was founded in 1817 by T.H. Gallaudet.

Death. Cessation of all life (metabolic) processes. Death may involve the organism as a whole (somatic death) or may be confined to cells and tissues within the organism. Somatic death has been traditionally defined as the discontinuance of cardiac activity and Respiration. However, with advances in organ Transplantation techniques, the need has arisen for a more precise definition of medical death, since the death of a donor must be established before transfer of an organ can be undertaken. Somatic death is followed by a number of irreversible changes of legal importance, especially in estimating time of death (e.g., rigor mortis, rigidity of the body caused by changes in muscle tissue, occurring 10 min. to several hours after death). The criterion used to established somatic death is still controversial, although a U.S. presidential commission in 1981 recommended that death be defined as "irreversible cessation of all functions of the entire brain, including the brain stem."

De Brogile Wave. A wave associated with a moving particle, having a wavelength (the *de Broglie wavelength*) given by $\lambda = h/mv$, where m is the particle's mass, v its velocity, and h the Planck constant. Under certain conditions particles of matter display wavelike behaviour—the wave character of the electron was first demonstrated in the Davisson-Germer experiment. The effect is not confined to elementary particles—diffraction of beams of atoms can also occur from crystal surfaces. [After Prince Louis Victor de Broglie (b. 1892), French physicist].

Debye. Symbol: D. A unit of electric dipole moment equal to $3.335\ 64 \times 10^{-30}$ coulomb metre. [After Peter Debye (1884-1966), Dutch physicist and physical chemist].

Deca. Symbol: Da. Prefix indicating 10.

Decahydrate. A solid compound with ten molecules of water of crystallization per molecule of compound, as in sodium sulphate decahydrate, $Na_2SO_4 \cdot 10H_2O$.

Decahydronaphthalene. A colourless flammable liquid, $C_{10}H_{18}$, with a menthol-like odour. It is a bicyclic hydrocarbon, made by hydrogenation of naphthalene, and is used as a solvent and fuel. M.pt. -43°C; b.pt. 195°C; r.d. 0.89.

Decay (Disintegration). Spontaneous disintegration of atomic nuclei: one nuclide (the *parent*) disintegrates yielding another type of nuclide (the daughter) and ejecting an electron or alpha particle and sometimes a gamma ray. The term also refers to the fall with time in the activity of a radioactive sample.

Decay Constant. *See* Radioactivity.

Deci. Symbol: d. Prefix indicating one tenth (10^{-1}).

Decibel. A measure of power level on a logarithmic scale. Although the fundamental unit is the *bel,* the decibel (0.1 bel) is almost always used. A given amount of power *P* has a power level in decibels of $10 \log_{10} (P/P_o)$, where P_o is a reference value. The decibel is used particularly for expressing the intensity of sound.

Decimal System. [Lat., = of tenths), Numeration system based on powers of 10 and using the digits 0,1,2,3,4,5,6,7,8 and 9. A Numeral in the decimal system is written as a row of digits, with each position in the row corresponding to a specific power of 10. A decimal point in the row divides it into increasing positive powers of 10 for positions to the left, and increasing negative powers of 10 for positions to the right. For example, 103 represents $(1 \times 10^2) + (0 \times 10^1) + (3 \times 10^0)$, or 100 + 0 + 3; and 1.03 corresponds to $(1 \times 10^0) + (0 \times 10^{-1}) + (3 \times 10^{-2})$, or 1 + 0 + 3/100. The decimal system was introduced into Europe c. 1300. Its positional system made it an improvement over Roman numerals and simplified arithmetic computation. The Metric System of weights and measures is based on the decimal system, as are most systems of national currency.

Declination (Variation). The angle made between the direction of the horizontal components of the earth's field at a point and the direction of true north at that point.

Decomposition. The breakdown of a chemical compound into simpler molecules.

Decompression Sickness. Physiological disorder caused by rapid decrease in atmospheric pressure that results in the release of nitrogen bubbles into body tissues. The bubbles cut off the oxygen supply, causing nausea, pain in the joints and abdomen, and, in severe cases, shock, paralysis, and even death. Also known as the bends, or altitude sickness, it affects divers, airplane pilots, and others working under compressed air. Use of a *decompression chamber* allows for a gradual reduction of pressure, preventing the accumulation of nitrogen in body tissues.

Decrepitation. The roasting of crystals until they emit a crackling sound or until this sound stops. The sound is produced by abrupt changes in crystal structure, usually due to the evolution of water of crystallization.

Deduction. 1. In traditional Logic, the process of drawing, by reasoning, particular conclusions from more general principles assumed to be true. The Aristotelian Syllogism is the classic example of deductive logic in the tradition.

2. In contemporary logic, any statement derived by a transformed rule upon an axiom; more generally, the term now refers to a process of deriving theorems from axioms, or conclusions from premises, by formal rules (transformation rules).

Deer. Ruminant Mammal of the family Cervidae, found worldwide except for Australia. Deer range in size from the Moose, which may be c.7 ft. (2.1 m) high at the shoulder, to the pudu of South America, which is c.12 in. (30 cm) high. Antlers, bony outgrowths of the skull used as weapons during the breeding season, develop in males of most species and are shed and renewed annually. In deer lacking antlers, long, upper canine teeth are used for fighting. The whitetailed deer, a source of food and buckskin for native Americans and white settlers, was nearly exterminated by over-hunting; it is now restored in the E U.S. Other North American deer include the mule deer of the the W U.S. and the closely related black-tailed deer of the Pacific Coast. Old World deer include the red deer and fallow deer.

Defect. A discontinuity in the regular structure of a crystal. A *point defect* is one caused by a vacancy or interstitial. A *line defect* or *dislocation* involves a number of atoms. A simple type occurs when part of one plane of atoms is missing from the lattice.

Defense Mechanism. In Psychoanalysis, any of a variety of unconscious reactions used by individuals to satisfy emotional needs: e.g., to establish harmony among conflicting strivings, to reduce feelings of Anxiety that result from unacceptable wishes, or to modify reality to make it more tolerable. Some common defense mechanisms are repression, the prevention of unacceptable ideas from entering the conscious mind, displacement, the release of dangerous impulses in a substitute situation or through disguised activity, and sublimation (a constructive mechanism), the redirection of instinctive sexual energy toward socially valuable ends.

Definite Proportions, law of. Constant composition.

Definition. The sharpness of the image formed by an optical system.

Degaussing. The proces of eliminating the magnetic field of an object, as by opposing it with an equal field produced by coils of wire.

Degenerate. 1. Denoting distinct energy levels in an atom, molecule, etc., that are equal in energy.

2. Denoting matter in stars in which the electrons have been removed from the atoms so that the material consists of a packed mass of electrons and atomic nuclei.

Degradation. A chemical reaction in which a compound is decomposed in several stages.

Degree. 1. A unit of plane angles, equal to 1/360 of a complete revolution.

2. An interval in certain scales of measurement, as in a temperature scale.

3. The sum of the exponents of the variables in a mathematical expression. For example $7x^6$ is of degree six, $5x^2yz^4$ is of degree seven. When an expression have several terms, its degree is that of its highest term; $9x^3 + 2x^2 + 3$ has degree three. It is also possible to speak of a degree in specified variables; x yz^3 is of degree seven in x, one in y, and three in z.

Degree of Freedom. 1. One of a number of modes by which an atom or molecule can take up energy. Any species has three translational degrees of freedom, corresponding to motion in three directions. In addition molecules have rotational and vibrational degrees of freedom. Linear molecules have two rotational degrees corresponding to rotation about two axes perpendicular to the axis (energy cannot be taken up by rotation about the axis). Nonlinear molecules have three degrees of freedom. The vibrational degrees of freedom correspond to vibrations between the atoms in the molecule and depend on the number of atoms.

2. The number of independent variables (temperature, pressure, etc.) that fix the state of a system.

Dehydration. 1. The removal of water from a mixture or from a compound containing water of crystallization.

2. The removal of hydrogen and oxygen from a compound in the ratio 2:1. Ethanol, for example, can be dehydrated to give ethylene: $C_2H_5OH = C_2H_4 + H_2O$.

Dekatron. A type of electron tube containing a low pressure of gas and ten small cathodes arranged in a circle about a central anode. The electron arrangement is such that only one cathode can function at a time and a glow discharge is transferred from one cathode to the next by a voltage pulse. Dekatrons are used as visible counters.

Deliquescence. Absorption of water by a hygroscopic solid resulting in the formation of a solution.

Delocalization. The existence in some molecules of electrons that are not confined to the region of one particular bond or atom but can move over the whole molecule. When this effect occurs the molecule is more stable than it would be if the electrons were localized.

Delta. The alluvial plain formed at the mouth of a river where the stream loses velocity and drops part of its sediment load. No delta is formed if the coast is sinking or if ocean or tidal currents prevent sediment deposition. A deltaic plain is usually very fertile but subject to floods. The three main varieties of delta are the arcuate (e.g., that of the Nile), the bird's-foot (e.g., that of the Mississippi), and the cuspate (e.g,, that of the Tiber).

Delta Iron. The allotropic form of iron stable above 1403°C. It has a body-centred cubic crystal structure.

Democritus. C.460-C.370 B.C., Greek philosopher. His atomic theory of the nature of the physical world, known to us through Aristotle's writings, was the most scientific theory proposed up to his time. He held that all living things are composed of tiny indivisible particles, called atoms, and that their constant motion explains the creation of the universe:

the heavier atoms clustered together to form the earth, while the lighter ones formed the heavenly bodies.

Demodulation. The process of separating the signal from a modulated carrier wave.

Demography. Science of human Population, especially of human numbers and their changes over time. By studying birth and death rates, marital and migration patterns, and population distribution, demography ascertains trends and predicts probabilities.

Denature. To render alcohol unfit to drink by the addition of a small amount of some other substance. Ethanol is usually denatured with poisonous methanol.

Dendrite. A crystal with a branching structure, giving it a tree-like appearance.

Dengue Fever. Acute infectious disease caused by a virus and transmitted by the *Aedes* mosquito. It occurs in warm climates. Symptoms include headache, fever, and intense joint pain, followed by a generalized rash. There is no specific treatment, but the disease can be controlled by eradicating the mosquitoes and their breeding grounds.

Denominator.. *See* Fraction.

Densitometer. An instrument for measuring the extent to which a material transmits or reflects light, usually using a standard light source and a photoelectric cell. Densitometers are used for measuring the amount of darkening of a photographic film and thus for finding the intensity of the radiation in spectrographic analysis, X-ray diffraction, and similar techniques.

Density. 1. Symbol: P. The mass per unit volume of a substance or material. Density is measured in kg $^{-3}$, g cc^{-3}, lb ft^{-3}, etc.

2. In general, any measure of the intensity of some property with respect to distance, area, or volume. For example, the current density at a point in a conductor is the electric current per normal unit area at that point. The charge density is the electric change per unit volume (volume charge density) or per unit area (surface charge density).

Density. Ratio of the Mass of a substance to its volume. Because many substances, especially gases, can be compressed into a smaller volume by increasing the pressure acting on them, the temperature and pressure at which the density is measured are usually specified.

Dentistry. Treatment and care of the Teeth and associated oral structures. Dentists are concerned with tooth decay, diseases of the supporting structures (such as the gums); faulty positioning of the teeth, and tooth replacements, as well as prevention of these problems. Specialized fields of dentistry include orthodontics (corrective dentistry) and periodontics (treatment of gum diseases.

Dependent Variable. *See* Variable.

Depolarization. The minimization of polarization effects in electrochemical cells by stirring the electrolyte, by increasing the operating temperature, or by use of a chemical reaction, as in the Leclanche cell.

Depression. A Disproportionately intense reaction to difficult life situations, characterized by strong feelings of loss, sadness, worthlessness, or rejection. It may be accompanied by such physiological symptoms as tense posture, persistent frowning, slowing of activity, fatigue, lack of appetite, and insomnia. The depressed individual's tendency toward self-recrimination, self-punishment, and guilt may ultimately lead to Suicide.

Depression of Freezing Point. A decrease in the freezing point of a solvent when a substance is dissolved in it. The depression is a colligative property: for pure substances it is proportional to the number of molecules or ions dissolved. Measurement of the elevation is a method of determining molecular weights.

Depth Charge. Explosive device used against submarines and other underwater targets. Delivered from surface ships, it is either rolled into the water or propelled by special throwers. The charge is detonated by water pressure at a predetermined depth and, if it explodes near enough, can destroy the target by concussion.

Des or **Diethylstilbestrol.** Synthetic nonsteriod female sex hormone having the same physiological effects as Estrogen. Effective even when taken orally, DES was formerly used in pregnant women to prevent miscarriage. During the 1970s the drug was linked to vaginal cancer in some women whose mothers had taken DES during pregnancy. The use of DES in a "morning after" contraceptive pill and as a growth accelerator in beef cattle is controversial because it may be a carcinogen.

Desalination. Process of removing soluble salts from water to render it suitable for drinking, irrigation, or industrial uses. In *distillation,* salt water is heated in one container to make the water evaporate, leaving the salt behind. The desalinated vapor is then condensed to form water in a separate container. Although long known, distillation has found limited application in water supply because of the fuel costs involved in vaporizing salt water. Other desalination techniques include *electrodialysis,* the use of porous membranes to filter out negatively and positively charged salt ions; *freezing,* based on the principle that water excludes salt when it crystallizes to ice; ion-exchange, in which water passes through a bed of specially treated synthetic resins that are capable of extracting ions of the salt from the solution and replacing them with ions that form water; and *reverse osmosis,* in which pressure, generated by the presence of salt in the water, forces water through a membrane permeable only pure water.

Descartes, Rene, 1596-1650. French philosopher, mathematician, and scientist. His philosophy is called Cartesianism (from *Cartesius,* the Latin form of his name). He is often called the father of modern philosophy, and his work is generally regarded as the bridge between Scholasticism and all philosophy that followed him. Primarily interested in mathematics, he was the founder of Analytic Geometry and originated the Cartesian Coordinates and Cartesian curves. To algebra he contributed the treatment of negative roots and the convention of exponent notation. Descartes also made important contributions to science, including works on optics, physiology, and psychology. His *Discourse on Method* (1637) and *Meditations* (1641) contain his important philosophical theories. Intending to extend mathematical method to all areas of human knowledge, Descartes discarded the authoritarian systems of the scholastic philosophers and began with universal doubt. Only one thing cannot be doubted: doubt itself. Therefore, the doubter must exist. This is the kernel of his famous assertion *Cogito, ergo sum* [I think, therefore I am]. From this kernel of certainly Descartes expanded knowledge, step by step, to admit the existence of God (as the first cause) and the reality of the physical world, which he held to be mechanistic and entirely divorced from the mind; the only connection between the two is the intervention of God. This is almost complete Dualism.

Desert. Arid region, usually partly covered by sand, supporting a limited and specially adapted plant and animal population. So called cold deserts, caused by extreme cold and often covered perpetually by snow or ice, form about one sixth of the earth's surface. Warm deserts form about one fifth of it. The world's largest desert areas lie between 20° and 30° north and south of the equator in regions where mountains intercept the paths of the trade winds or where atmospheric conditions limit precipitation. An area with an annual rainfall of 10 in. (25 cm) or less is considered a desert, although some deserts and semideserts exist in areas of higher precipitation where moisture is lost by run-off or evaporation.

The two largest deserts in the world are the Sahara in Africa and the great desert of central and W. Australia.

Desiccator. A device for drying or for keeping materials dry. The common form is an air-tight compartment, which may be evacuated, in which the material is placed on a mesh stand, below which is a hygroscopic substance such as phosphorus pentoxide or calcium oxide.

Desorption. The opposite process to adsorption: i.e. the removal of molecules, ions, etc., from a surface into the gas phase.

Destructive Distillation. A process in which a substance or mixture of substances is heated leading to decomposition and formation of volatile products, which can be collected. For example, the destructive distillation of wood produces methyl alcohol, the destructive distillation of coal produces coal gas and coal tar.

Determinant. A mathematical expression consisting of a square array of numbers or variables, used in the solution of simultaneous equations.

Detector. 1. Any device or instrument showing the presence of or for measuring some effect or physical property, such as streams of particles, electromagnetic radiation, heat, etc.

2. The circuit used to separate the signal from a carrier wave.

Detergent. Any substance that can be used as a cleansing agent. The term is usually applied to substances that resemble soap in their action but are not made by saponification of fats or oils. Like soap, detergents are surfactants which have a hydrophobic (oil-soluble) and a hydrophilic (water-soluble) part in their molecules. The hydrophobic portion is usually a long hydrocarbon chain. There are three types of detergent depending on the nature of the hydrophilic part. The most important, *anionic detergents,* are salts of organic acids: the negative ion consists of a hydrocarbon chain attached to a negative (hydrophilic group). Commonly anionic detergents are salts of sulphonic acids, containing ions of the type RSO_2O. In *cationic detergents* (or *invert soaps*) the active agent is a positive ion. Usually these compounds are quaternary ammonium salts of simple acids. The cation has the form RNH_3^+, in which R is a long hydrocarbon chain (the hydrophobic part) and $-NH_3^+$ is the hydrophilic part: the hydrogen atoms may be replaced by small organic groups. *Nonionic detergents* have neutral molecules in which the hydrophobic part is a polar group, usually containing oxygen atoms.

Synthetic detergents are extensively used as household and industrial cleansing agents and in processes requiring surfactants. Unlike soap, they are unaffected by hard water. Household detergents contain other substances such as perfume, whitening agents.

Detonation. An explosion often a small explosion used to initiate a second, larger one.

Deuterated. Indicating that a chemical compound contains deuterium atoms in place of hydrogen atoms, as in deuterated methane, CD_4.

Deuterium (Heavy Hydrogen). Symbol: D, ^2H. The hydrogen isotope with one proton and one neutron in its atomic nucleus. It occurs naturally with the normal isotope (0.0156%) and is prepared by the electrolysis of water. The properties are almost identical to those of normal hydrogen. It is used as a tracer in chemical reaction studies. A.N.1; A.W. 2.014.

Deuterium Oxide. Heavy Water.

Deuteron. A nucleus of a deuterium atom.

Deviation. The difference between a particular value of an observation and either the true value or the average value of a number of observations. The *mean deviation* is the average of the absolute (positive) values of a set of deviations.

Devitrification. The destruction of the amorphous structure of a glass, by crystallization. It results in brittleness and loss of transparency. Repeated heating and cooling of the glass eventually causes devitrification but the most common cause is the presence of foreign material, such as carbon, when the glass is in a molten state.

De Vries, Hugo. 1848-1935, Dutch botanist. He rediscovered (1900) Gregor Mendel's laws of heredity and developed the theory of Mutation, expounded in *The Mutation Theory (1901-3)* and in *Plant-Breeding* (1907). De Vries introduced the experimental method to the study of evolution, and maintained that new species develop via mutations and that each equality subject to change is represented by a single physical unit (which he called a pangen). He work on osmosis is also important; he coined the term *isotonic.*

Dew. Thin film of water that has condensed on the surface of objects near the ground. Dew forms when objects cool during the night, causing the layer of air in contact with them to cool, and then causing the condensation of the water vapor in that layer of air, Condensation occurs because the capacity of air to hold water vapor decreases as the air is cooled. Frost forms if the dew point, or temperature at which condensation begins, is below 32°F (0°C).

Dewar flask (vacuum flask). A vessel for maintaining its contents at a temperature above or below that of the surrounding air, having double walls with the space between them evacuated to prevent loss of heat by conduction. Usually the flask is made of thin glass, silvered on the inside surface to prevent loss of heat by radiation. [After Sir James Dewar (1842-1923), Scottish chemist].

Dewar, Sir James. (1842-1923), British chemist and physicist; b. Scotland. Professor at the Royal Inst., London, and later director of the Davy-Faraday Research Laboratory there, he was knighted in 1904. Best known for his work on the properties of matter at very low temperatures and the liquefaction of gases, he liquefied and solidified hydrogen and invented the Dewar Flask.

Dew point. The temperature at which water vapour begins to condense from humid air as the air is cooled. It is the temperature at which the air is saturated and depends on the humidity.

Dextrorotatory (dextrorotary). Denoting a chemical compound that is capable of imparting a clockwise rotation to the plane of polarization of polarized light, as observed by someone facing the light's direction of motion.

Dextrose. *See* glucose.

Diabetes or **Diabetes mellitus.** Chronic disorder involving abnormal Insulin secretion, associated with high blood sugar (hyperglycemia) and altered fat and protein metabolism. Diabetics are predisposed to developing vascular disease (e.g., Arteriosclerosis), kidney disease, Hypertension, and Blindness. Diabetics who secrete insufficient amounts of insulin are known as non-insulin-dependent diabetics. This type of diabetes, most commonly seen in overweight adults, can often be managed by weight loss and diet regulation although sometimes insulin or oral hypoglycemic drugs may be needed. Diabetics unable to produce their own insulin are known as insulin-dependent and must receive injections of the hormone daily.

Diacetyl. A yellow liquid ketone, $CH_3COCOCH_3$, which smells of butter. It is used as a flavouring. M.pt.—4°C; b.pt. 88° C; rd. 0.99.

Dialysis. In chemistry, transfer of dissolved solids (solute) across a semipermeable membrane, which permits or hinders diffusion of molecules according to their size. Dialysis is frequently used to separate different components of a solution. For example, in artificial kidney machines dialysis is used to purify the blood of persons whose Kidneys have ceased to function. In the machine, blood is circulated on one side of a semipermeable membrane, while dialysis fluid—containing substances necessary to the body and closely matching the chemical composition of the blood—is circulated on the other side. Metabolic waste products, such as urea, diffuse through the membrane into the dialysis fluid and are discarded, while the diffusion of substances necessary to the body is prevented. *See* also Osmosis.

Diamagnetism. A type of magnetic behaviour in which the material has a low negative susceptibility that is independent of temperature. A diamagnetic sample will tend to move away from an applied magnetic field. The effect is due to the orbital motion of electrons in the atom or molecule. When a field is applied to the sample the electrons change their orbit so that they tend to resist the application of the field by producing an opposing field. All substances exhibit diamagnetism but this is often swamped by paramagnetic or ferromagnetic behaviour.

Diaminoethane (ethylene diamine). A colourless viscous liquid amine, $NH_2CH_2CH_2NH_2$. It is prepared by reacting dischlorethane with ammonia and is a solvent, stabilizer for rubber latex, and corrosion inhibitor. M.pt. 8.5°C; b.pt. 116° C; r.d. 0.89.

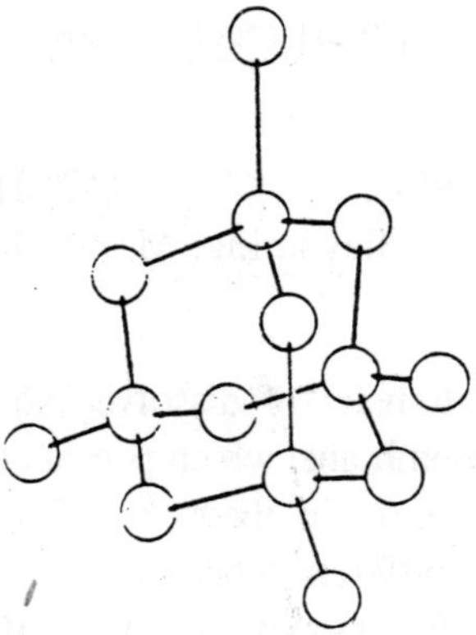

Fig. D-1. Diamond structure.

Diamond. 1. A naturally occurring allotrope of carbon. It is an extremely hard crystalline substance. When pure it is transparent and has a high refractive index and dispersive power. Coloured highly impure varieties are used for drills and other cutting tools. The substance owes its hardness to its covalent crysal structure.

2. Mineral, one of two crytalline forms of the element Carbon. It is the hardest substance known, and inferior stones are used as Abrasives, in certain types of cutting tools, and as phonograph needles. Gem diamonds were first found in streambeds in India and Berneo; most now come from volcanic pipes in South Africa. Famous diamonds include the Koh-i-noor, now among the English crown jewels; the Cullinan, from which 105 stones were cut; and the blue Hope diamond. Synthetic diamonds, produced since 1955, are now widely used in industry.

Diamond-ring effect. The phenomenon occurring just after totality in a total solar eclipse when the dazzling face of the sun (the photosphere) becomes visible from behind the body of the moon and before the corona has had time to fade.

Diastereoisomer. *See* optical isomerism.

Diathermancy. The property of transmitting infrared radiation. *Compare* athermancy.

Diatom. Single-celled, macroscopic plant that secretes and is enclosed by an often intricate, round-to-elongated silica shell. Golden-brown Algae, diatoms are found in fresh and salt water, in moist soil, and on the moist surfaces of other plants; they are the principal constituent of Plankton, an important food source for aquatic animals. Most exist singly, but some from colonies. When the aquatic forms die, their shells collect in the ooze on the bottom, eventually forming the material called diatomaceous earth (kieselguhr) or the more compact, chalky, light-weight rock called diatomite, used in sound and heat insulation, in making explosives, and for filters and abrasives. Most Limestone and much petroleum is of diatom origin.

Diatomic. Having molecules composed of two atoms. Hydrogen, H_2, and hydrogen chloride, HCl, are both examples of diatomic gases.

Diazine. Any of three isomeric azines, $C_4N_2H_4$. The isomers are pyrazine, pyrimidine, and pyridazine.

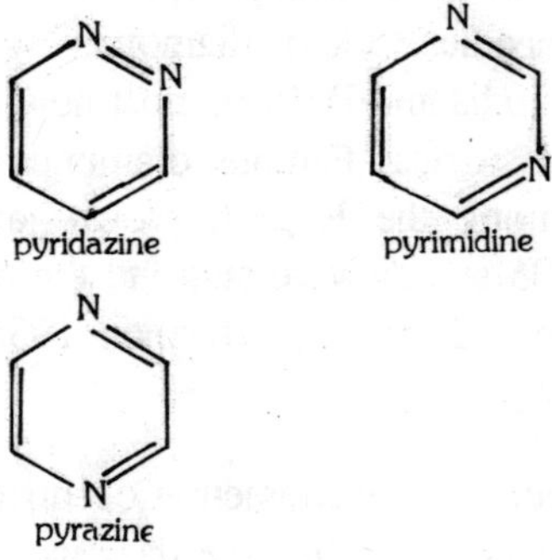

Fig. D-2. Diazines.

Diazo compound. Any of a class of compounds of the formula R:N:N. where R is an alkyl group. The term is also used for azo compounds, which contains the group -N:N-, and for diazonium salts.

Diazonium salt. Any of a class of salts with the general formula $ArN_2^+X^-$, where Ar is an aryl group. Diazonium salts are made by using an amine dissolved in an acid and adding sodium nitrite to form nitrous acid, which reacts with the amine. They are intermediates in making azo compounds.

Dibromoethane (ethylene dibromide). A colourless oily liquid, $BrCH_2CH_2Br$, made by the addition of bromine to ethane. It is used as a solvent. M.pt. 9°C; b.pt. 121°C; r.d. 2.2.

Dacarboxylic acid. An acid that has molecules containing two carboxyl groups.

Dice. [Plural or *die*], small cubes used in games, usually made or ivory, plastic, or similar materials. The six sides are numbered by dots from 1 to 6, placed in such a way that the sum of the dots on opposite sides equals 7. Dice games have been popular since ancient times. In the simplest form of play, each player throws, or shoots, for the highest sum. In craps, a popular gambling game in the U.S., the "shooter" rolls two dice, winning on the first roll if 7 or 11 is thrown and losing if the sum is 2, 3, or 12. If any other number is thrown, the player continues to shoot until that number is duplicated, for a win, or a 7 is rolled, for a loss.

Dichlorobenzene. Any of three isomeric compounds, $C_6H_4Cl_2$, prepared by the direct catalytic chlorination of benzene. The most widely used isomer is *1,4-dichlorobenzene,* a white crystalline solid used as a moth repellent and deodorant. M.pt. 53.1°C; b.pt. 174°C; r.d. 1.28. The other two isomers are colourless volatile liquids: *1, 2,-dichloro-benzene* is used as an insecticide and an intermediate for making dyes. M.pt. -17°C; b.pt. 180°C; r.d. 1.30.

Dichlorodifluoromethane. A colourless odourless fairly inert nonflammable gas, CCl_2F_2, manufactured by reacting carbon tetrachloride with hydrogen flouride and used as a refrigerant, aerosol propellant, and solvent.

Dhchloroethane (ethylene dichloride). A colourless odourless flammable liquid. $ClCH_2CH_2Cl$, made by the addition of chlorine to ethene. It is used in making vinyl chloride and as a solvent. M.pt. — 35°C; b.pt. 83°C; r.d. 1.3 (20°C).

Dichloromethane (methylene chloride). A colourless volatile liquid, CH_2Cl_2, prepared by the chlorination of chloromethane. It is used as a solvent and as a local anaesthetic. M.pt. -97°C; b.pt. 40°C; r.d. 1.3.

Dichromate. A salt containing the divalent ion $Cr_2O_7^{2-}$. Dichromates are salts of the hypothetical *dichromic acid,* $H_2Cr_2O_7$.

Dichromic acid. *See* dichromate.

Dielectric. 1. A non-conductor of electricity, especially a substance with an electrical conductivity of less than 10^{-6} siemen.

2. Material that does not readily conduct electricity, i.e., an insulator (see Insulation). A good dielectric resists breakdown under high voltages, does not draw appreciable power from the circuit, and has reasonable physical stability. Dielectrics are used to separate the plates of a Capacitor. The dielectric strength is a measure of the maximum voltage that a dielectric can sustain without significant Conduction

Dielectric constant. *See* relative permitivity.

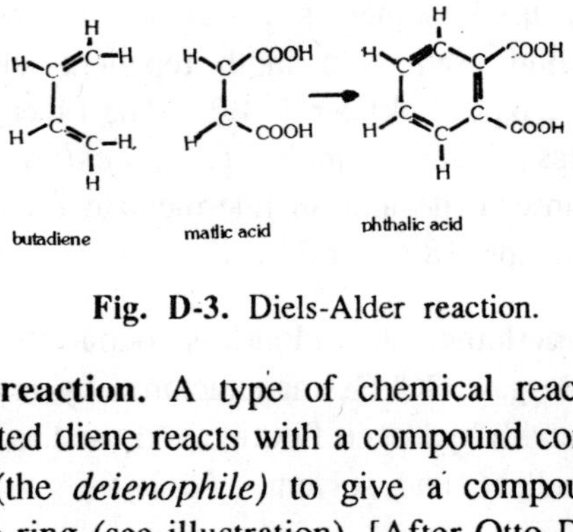

Fig. D-3. Diels-Alder reaction.

Diels-Alder reaction. A type of chemical reaction in which a conjugated diene reacts with a compound containing a double bound (the *deienophile)* to give a compound containing a benzene ring (see illustration). [After Otto Diels (1876-1954) and Kurt Alder (1902-1958), German chemists.]

Diene. An organic compound containing two carbon-carbon-double bonds in its molecules.

Dienophile. *See* Diels-Alder reaction.

Diesel Engine. Type of Internal-Combustion Engine patented (1982) by the German engineer Rudolph Diesel. The diesel is heavier and more powerful than the gasoline engine and burns fuel oil instead of gasoline. It differs from the gasoline engine in that the ignition of fuel is caused by compression of air in its cylinders instead of by a spark. The speed and power of the diesel are controlled by varying the amount of fuel injected into the cylinder—not the amount of air admitted, as in the gasoline engine. Diesels are widely utilized to power industrial and municipal electric generators, continuously operating pumps such as those used in oil pipelines, and ships, trucks, locomotives, and some automobiles.

Diethylamine. A colourless flammable liquid amine, $(C_2H_5)_2NH$, with an odour resembling that of ammonia. It is prepared by reacting chloroethane with ammonia and is used in the rubber and petroleum industries and in the manufacture of dyes and pharmaceuticals. M.pt. -50°C; b. pt. 55°C; r.d. 0.71.

Diethylene glycol. A colourless odourless hygroscopic syrupy liquid, $CH_2OH\text{-}CH_2OCH_2CH_2OH$, manufactured by heating ethylene oxide with ethylene glycol. It is used as a solvent, antifreeze, and drying agent. M.pt. -8°c; b.pt. 245°C; r.d. 1.12.

Diethyl ether (diethyl oxide, ether). A very light volatile colourless flammable liquid. $(C_2H_5)_2\text{-}O$. manufactured by the reaction between sulphuric acid and ethanol. It iş used as an anaesthetic and solvent. M.pt. -116°C; b.pt. 34.5°C; r.d. 71.

Diethyl oxide. *See* diethyl ether.

Differential calculus. The branch of calculus concerned with differentiation.

Differential equation. An equation containing derivatives. The order of the highest derivative is the order of the equation; for example $d^2y/dx^2 = -ax$ is a second-order differential equation.

Differentiation. A mathematical process concerned with finding the way in which a function changes with respect to one of its variables. If y is a function of x. written y = f(x), the *derivative* of *y* with respect to *x* is the limit of a small change in *y* divided by a small change in *x* as the change in *x* becomes infinitesimally small; i.e. the limit of $\Delta y/\Delta x$ as Δx tends to zero. In general the derivative, which is written *dy/dx,* Dy, or *y*, will also be a function of *x*. The derivative of a function at a particular value of *x* is the gradient of a curve of the function at that point. The differentiation of distance with respect to time gives velocity. Further differentiation can be performed. The second derivative is written d^2y/dx^2, etc. The number of differentiations is the *order* of the derivative. *See also* partial derivative.

Diffraction. 1. The formation of light and dark bands *(diffraction patterns)* around the boundary of a shadow cast by an object or aperture. The effect results from interference of the light from the edges of the object.

2. Interference of waves reflected or scattered from an array of lines, atoms, etc. *See* diffraction grating, X-ray diffraction, electron diffraction.

Diffraction grating. A glass plate or metal mirror ruled with large numbers of closely spaced parallel lines. If the line separation is comparable with the wavelength of light the grating reflects or transmit light with interference so that a series of widely spaced bright lines are produced. The condition for such a maximum is $d(\sin i + \sin \theta) = n\lambda$, where *d* is the spacing between lines on the grating, *i* the angle of incidence. θ the angle the reflected or transmitted light makes with the normal, and *n* is an integer. 0, 1, 2, etc. If $n = 1$, the value θ for a given value of *i* depends on the wavelength. Thus polychromatic light is separated into a spectrum by the grating. Higher-order spectra are obtained for higher values of *n*. Diffraction gratings are used in spectrometers for dispersing light and ultraviolet radiation.

Diffuse reflection. Reflection in which the angle of reflection does not equal the angle of incidence. It is distinguished from specular reflection and occurs when the reflecting surface is uneven, surface irregularities being comparable in size to the wavelength of the reflected radiation.

Diffusion. 1. Motion of a substance as a result of the random thermal motion of its atoms or molecules. Two liquids or gases place in contact will slowly become completely mixed as a result of this process. Gasses obey Graham's law of diffusion: the rate of diffusion is inversely proportional to the density. Thus heavy molecules diffuse more slowly than light molecules, a fact exploited in the separation of isotopes of uranium by diffusion of the volatile hexafluoride through a porous barrier. Solids also undergo diffusion (as can be demonstrated by tracer studies) but the process is very slow at normal temperatures.

2. Irregular reflection or transmission of light involving scattering.

Diffusion pump. A type of vacuum pump in which gas is removed from a system by a steam of mercury or oil vapour. The vapour issues from a nozzle and molecules of gas in the system diffuse into this jet and are carried away. The vapour is condensed inside the pump. Diffusion pumps can only operate at low pressures and are usually backed by a mechanical pump.

Digestive system. In animals, a group of organs that digest food and eleminate wastes. Digestion converts food into a form that can be absorbed into the Circulatory System and distributed to tissues throughout the body. Digestion begins at the mouth, where chewing reduces the food to a fine texture, saliva moistens it, and an enzyme (amylase) in saliva begins the conversion of starch into simple sugars. The swallowed food passes through the Pharynx and through a long tube, the esophagus, into the Stomach. The food is pushed through the digestive system by peristalsis (contractions of the muscles forming the outer walls of the digestive tract). In the stomach, where digestive Enzymes and gastric juices are secreted, sugars and alcohol are absorbed directly through the stomach wall into the blood stream, and the remainder, in the form of a thick liquid (chyme), passes into the first section of the small Instestine the duodenum). There digestive enzymes from the Pancreas break down fats into glycerol and fatty acids, starches into sugars, and proteins into amino acids. The Liver supplies bile, which makes fat globules smaller and more easily digested. Small glands in the intestinal wall (crypts of Lie berkuhn) also secrete enzymes that continue digestion. Digested food is absorbed through small projections of the intestinal wall (villi) into the blood and lymph. Undigested material passes into the large Instestine and is excreted through the anus.

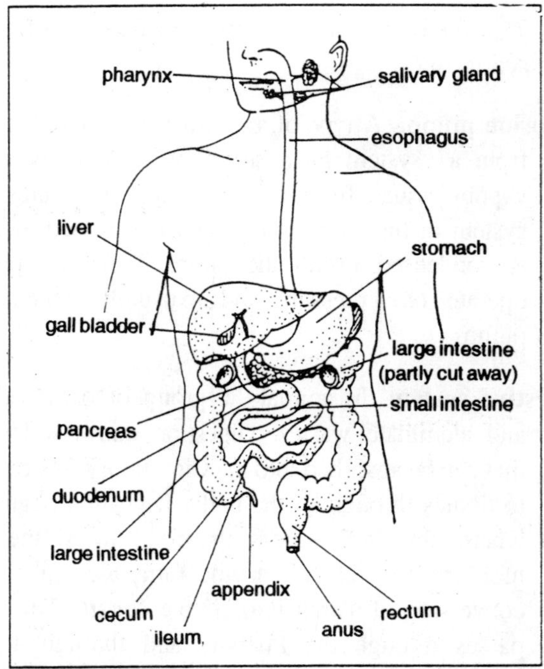

Fig. D-4. Digestine system.

Digit. Any of the symbols used in a notation for the integers. The number of different digits used in such a system is the *base* of the system. The binary system has two digits and the decimal system has ten.

Digital circuit. Electric Circuit in which the output currents or voltages are interpreted as having one of several (often two) values, depending on which a corresponding number or ranges they fall into. Such circuits are used to implement logical operations or operations on representations of discrete numbers, often in binary form. *See also* Analog Circuit.

Digital computer. A type of computer in which information is stored and handled in the form of discrete units. Modern computer systems are often extremely complex and sophisticated pieces of equipment, but all consist of a few basic units.

The input system is the part used to feed instructions and data into the device: the instructions are usually written in the form of a programming language. Here the information is converted into the form in which it is stored and handled. In nearly all digital computers some type of binary notation is used. Thus, numbers, letters, and other symbols are represented by a string of binary digits (0 and 1), each of which can be represented within the computer by one of two alternative states of a component; i.e., by the presence or absence of a current pulse, voltage, or magnetized region. The information, including the program and the data, is then held in the computer store.

Operations Performed on the data are carried out by electronic circuits that form the central processing unit. Basically, these contain combinations of switching circuits (gates), which manipulate data drawn from the store in the form of a string of pulses. Thus, combinations of gates are capable of adding subtracting, and performing simple logical operations such as comparison. The resulting information is again stored until it can be passed to the output system, where it is converted back into usable form and printed, displayed on a screen, etc.

Within the computer operations are governed by the *control,* the electronic circuitry that controls transfer of data from input to store, store to central processing unit, etc.

The characteristic features of digital computers are their ability to store information and their ability to perform operations according to a present program. They are capable of performing complex calculations in very short times and are also used extensively for information processing; i.e., for sorting vast amounts of information, storing and recovery of data, etc. *See also* analog computer, hybrid computer.

Digitalin. A white powder, $C_{36}H_{56}O_{14}$, extracted from the leaves and seeds of digitalis (foxglove) and used as a cardiac stimulant. M.pt, 210°C.

Digitalis. Any of several chemically similar drugs, including digitoxin, digoxin, and ouabain, used to treat heart failure are cardiac arrhythmias. Digitalis. Digitalis slows the pulse and increases the force of heart contractions and the amount of blood pumped per heartbeat. The chief source of the drug is the foxglove plant, *Digitalis purpurea.*

Digital-to-analog conversion. The process of changing discrete data into a continuously varying signal. The most common use is to present the output of a digital Computer as a graphic display (see Computer Graphics) or as audio output, as in computer-generated music. *See also* Analog-to-Digital Conversion.

Digital voltmeter. A type of voltmeter in which the reading is given in digital form rather than by a pointer on a scale. Digital voltmeters usually have a solid-state amplifier producing a steady voltage, which is compared with a reference voltage inside the instrument. The fererence voltage increases linearly and at the same time pulses are generated at a constant rate and counted. When the reference voltage reaches the measured voltage the pulses stop, the whole process being repeated every few seconds.

Dihydrate. A solid compounds with two molecules of water of crystallization per molecules of compound, as in sodium bromide dihydrate, $NaBr,2H_2O$.

Dihydric. Denoting an alcohol or phenol containing two hydroxyl groups per molecule.

Dilatancy. A phenomenon shown by some non-Newtonian fluids in which the viscosity increases as the rate of shear increases. In other words, the fluid becomes more viscous the faster it moves. The effect is observed in some suspensions and pastes: it is caused by friction between the suspended particles. The opposite effect, thixotropy, is more common.

Dilatometer. Any apparatus suitable for measuring changes in volume: typically a glass bulb with a narrow capillary tube attached so that expansion or contraction of a liquid in the apparatus can be measured by the distance moved by the liquid's meniscus in the capillary tube.

Dill. Annual or biennial plant *(Anethum graveolens)* of the Carrot family, native to Europe. Its pungent, aromatic leaves and seeds are used for pickling and for flavoring sauces, salads, and soups.

Dilution. The process of reducing the concentration of a solute by addition of solvent.

Dimension. In mathematics, number of parameters or coordinates required to describe points in a mathematical object (usually geometric). The space we inhabit, having height, width, and depth, is three-dimensional; a plane or surface is two-dimensional; a line or curve is one-dimensional; and a point is zero-dimensional. By means of a coordinate system, e.g., Cartesian Coordinates, one can specify any point with respect to a chosen origin (and coordinate axes through the origin, in the case of two or three dimensions). By analogy, an ordered set of four, five, or more numbers is defined as representing a point in a space of four, five, or more dimensions.

Dimensions. The set of powers of basic independent physical quantities in terms of which a given physical property is defined. Many quantities can be expressed in terms of mass *(M),* length *(L),* and time *(T).* For example, area has dimensions of L^2 density has dimensions of ML^{-3}, force has dimensions of MLT^{-2}, etc. *Dimensional analysis* is a method of testing the correctness or partially deriving equations by use of dimensions. It depends on the fact that each term of an equation must have the same dimensional formula. Thus, in Einstein's equation, $E = mc^2$, both side of the equation have dimensions of ML^2T^{-2}.

Dimer. A compound whose molecules are formed by addition of two molecules of a simpler compound (the *monomer).* A reaction leading to formation of a dimer is a *dimerization.* Such reactions are often reversible, leading to equilibium between the monomer and the dimer. An example is the dimerization of nitrogen dioxide : $2NO_2 = N_2O_4$.

Dimethyl dichlorosilane. A colourless flammable liquid, $(CH_3)_2SiCl_2$, manufactured by reacting silicon with chloromethane. It is used in producing silicone oils, rubbers, and resins. M.pt. M.pt. -86°C; b.pt. 70°C; r.d. 1.06.

Dimethyl ether (methyl ether, wood ether). A colourless flammable gas, CH_3OCH_3, prepared by the dehydration of methanol. It is used as a refrigerant, solvent, and propellant for sprays. M.pt. -141°C; b.pt. 24°C; r.d. 0.66.

Dimorphism. The existence of two solid (crystalline) forms of the same chemical substance. Dimorphism in elements is commonly referred to as allotropy. Some dimorphs exist as modifications in the same crystal system, as in ammonium chloride where both forms are cubic, or they may crystallize in different crystal systems, as in hexagonal and cubic silver iodide. Each dimorphic form is stable within a particular temperature and pressure range and, at a given pressure, the temperature at which one form becomes the other is fixed and called the *transition point.* Existence of more than two forms is *polymorphism.*

Dinitrobenzene. Any of three white crystalline isomeric compounds with the formula, $C_6H_4(NO_2)_2$. Direct nitration of benzene gives *1, 3-dinitrobenzene,* which is used in the production of dyes and celluloid. M.pt. 90°C; b.pt. 303°C; r.d. 1.58. The 1, 2-and 1.4- isomers are prepared by oxidation of the appropriate nitroaniline.

Dinosaur. Any of a large group of extinct Reptiles (subclass Archosauria) that dominated the earth for nearly 150 million years in the Mesozoic era. Dinosaurs ranged in length from the 2-ft (60-cm) *Compsognathus* to the nearly 90-ft (27-m) *Diplodocus,* and in weight up to the 75-ton *Brachiosaurus.* They are traditionally classified as cold-blooded reptiles, but recent evidence on posture, skeleton, and eating habits suggests some may have been warm-blooded. They are all thought to have been egg-layers. Saurischian (order Saurischia), or lizard-hipped, dinosaurs included meat-eating bipeds, e.g., *Allosaurus* and *Tyrannosaurus,* and plant-eating quadrupeds, e.g., *Brontosaurus* and *Diplodocus.* Ornithischian (order Ornithischia), or bird-hipped, dinosaurs, included plant eating bipeds, e.g., *Iguanodon* and the duckbills, and armored dinosaurs, e.g., the plated *Stegosaurus* and horned *Triceratops.* It is not known why dinosaurs died out. Most theories postulate that changes in geography, climate, and sea level were responsible, and one recent theory suggests that an asteroid impact caused the changes that led to their extinction.

Diode. 1. An electronic component consisting of a semiconductor junction with two attached electrodes. The p-n junction acts as a rectifier: if a positive voltage is applied to the p-region a current flows whereas very little current flow occurs in the opposite direction.

2. A thermionic valve with two electrodes, an anode and cathode. Current flows through the valve when the anode has a positive potential with respect to the cathode but no electrons flow when it is negative with respect to the cathode. Diode valves can be used as rectifiers.

3. Two-terminal device having a low Resistance to electric current in one direction and a high resistance in the reverse direction. Diodes are thus useful as Rectifiers, converting alternating current (AC) into direct current (DC). Although Electron-Tube diodes were once common, almost all diodes today are Semiconductor devices. In general, current flowing through a diode is not proportional to the voltage between its terminals. When the voltage applied in the reverse direction exceeds a certain value, a semiconductor diode breaks down and conducts heavily in the direction of normally high resistance. This effect can be exploited to regulate voltage. Some diodes are sensitive to light (see Photoelectric cell, Photovoltaic cell). A lightemitting diode (LED) produces light as current passes through it; some LEDs can act as lasers. A thermistor is a special semiconductor diode whose conductivity increases with the diode temperature.

Diol (glycol. A dihydric alcohol; an alcohol containing two hydroxyl groups per molecule.

Dioptre. A unit of power of a lens, equal to the power of a lens with a focal length of one metre. The power in dioptres is the reciprocal of the focal length in metres. By convention, a converging lens has a positive power and a diverging lens has a negative power.

Dip (inclination). The angle that the earth's magnetic field makes with the horizontal at a point on the earth's surface.

Dip circle (inclinometer). An instrument for measuring magnetic dip, consisting of a magnetized needle pivoted to swing in a vertical plane over an angular scale.

Diphenylamine. (DPA). A colourless crystalline solid amine.

$(C_6H_5)_2NH$, prepared by heating aniline with anilinium chloride. Diphenylamine is used in the manufacture of dyes and stabilizers M.pt. 53°C; b.pt. 302°C; r.d. 1.16.

Diphosphate. *See* phosphate.

Diphosphoric acid. *See* phosphoric acid.

Diphtheria. Acute, contagious disease caused by the *Coryne bacterium diphtheriae.* It is spread through respiratory droplets of infected individuals. The bacteria, lodging in the mucous membranes of the throat, secrete a potent toxin, which causes tissue destruction and the formation of a gray membrane in the upper respiratory tract that can loosen and cause asphyxiation. The toxin may also spread via the blood and damage tissues elsewherein the body. Diphtheria can be prevented by vaccination.

Dipole. The electric charges of equal magnitude and opposite sign a small distance apart. The *dipole moment* is the product of the magnitude of one charge multiplied by the distance between the charges. Molecules can have permanent dipole moments when there is unequal sharing of the electrons in the bond between two atoms. *See also* magnetic dipole.

Dipole aerial. *See* aerial.

Dirac, Paul Adrien Maurice. 1902—, English physicist. He formulated (1928) a version of quantum mechanics that took into account the theory of Relativity. This theory implied the existence of an antiparticle to the electron; it was discovered later and named the positron. His equation of a particle's motion is a relativistic modification of Erwin Schrodinger's basic equation of quantum mechanics. Dirac and Schrodinger shared the 1933 Nobel Prize in physics. Dirac also helped to formulate the Fermi-Dirac statistics and contributed to the quantum theory of electromagnetic radiation.

Direct current (d.c.). An electric current that has a fairly constant value and always follows in the same direction.

Direct day. A dye that is attached directly to its substate; i.e., without the use of mordant.

Disaccharide. A sugar with molecules containing two monosaccharide units.

Discharge. 1. Removal of electric charge, as from a capacitor or cell.

2. Conduction of electricity through a gas or other insulating material. A simple gas discharge can be produced by placing two metal electrodes in a gas with a potential difference between them. Under normal conditions conduction does not occur because the gas molecules are neutral. If an external source of ionizing particles or radiation is available gas molecules can be ionized: the electrons move towards the anode and the ions move towards the cathode. Thus producing a flow of electric current. If the electric field between the electrodes is high the electrons may gain enough kinetic energy to ionize other molecules and large numbers of ions and electrons can be formed. When this occurs the discharge can become self sustaining and independent of the external source of ionization. Usually the discharge is initiated spontaneously by cosmic rays or field emission at a high potential difference. The voltage required to maintain the discharge is much lower.

In general, discharges in gases consist of a luminous plasma containing ions, electrons, free radicals, and excited atoms and molecules in addition to the normal molecules. The colour of the plasma depends on the nature of the gas. A complex set of processes is occurring, including ionization, recombination of ions and electrons, secondary emission from the electrodes, and excitation of molecules by impact of electrons. The precise behaviour depends on a variety of factors such as the pressure of gas, the potential difference and the current in the circuit, and the sizes, shapes, materials, and separation of the electrodes (Townsend discharge, arc, spark, corona discharge, brush discharge).

Discharges in gases are usually produced in glass or quartz discharge tubes containing gas at low pressures and electrodes sealed through the tube. Discharges can also be produced without electrodes by applying a radio- frequency field. The plasma is formed by acceleration of electrons in the gas. Discharge tubes are extensively used in studies of plasmas and as sources of light and ultraviolet radiation.

Disintetgration. *See* decay.

Disk (Disc). A direct-access storage device used in computers. consisting of a flat circular plate coated with a layer of magnetic iron oxide. The disk rotates on a spindle and the data is stored on a series of concentric tracks access being made by means of a head moved radially over the disk's surface. *See* store.

Dispersion. 1. The separation of a complex wave into component parts according to some characteristic such as frequency or wavelength, as in the separation of visible radiation into colour components by diffraction. The term particularly used for the way in which a transparent medium can disperse polychromatic radiation as a result of the dependence of refractive index on wavelength. The *dispersive power (w)* of a medium is defined by the equation $w = n_1 - n_2)/(n - 1)$, where n_1 *and* n_2 are the refractive indices of the medium for two specified wavelengths and n is the mean refractive index.

2. A suspension of solid, liquid, or gaseous particles of colloidal size or larger, in a homogeneous medium. The medium forms the *continuous phase* and the suspended material is the *disperse phase.*

Dispersion force. A very weak force existing between atoms and molecules. Dispersion forces result from mutual interaction between the electron-nucleus dipoles of two different atoms.

Dispersive power. *See* dispersion.

Displacement. 1. A chemical reaction in which one atom or group replace another in the molecule. A typical example is the hydrolysis of an alkyl halide, $H_2O + CH_3Cl = HCl + CH_3OH$, in which the chlorine atom is replaced by a hydroxyl group.

2. Symbol : **D.** A measure of the electric flux density in a medium, given by eE, where E is the permitivity and **E** is the electric field strength in free space.

Disproportionation. A chemical change involving simultaneous reduction and oxidation of the same compound, as in the reaction $2CuCl = Cu + CuCl_2$.

Dissociation. The breakdown of a molecule into radicals, ions, atoms, or simpler molecules. The term is often used to denote reversible decompositions, such as the ionization of acids in aqueous solution.

Dissociation constant. The equilibrium constant of a dissociation reaction. Acid dissociation constants are the equilibrium constants for reactions of the type : $HA + H_2O = H_3O^- + A^-$. Since the concentration of water is unity, the dissociation constant *K* is equal to $[H_3O^+]$ [A]/[HA]. *See also* pK.

Distillation. 1. A process used to purify or separate liquids by evaporating them and recondensing the vapour. The liquid sample is boiled in a flask attached to a condenser. In this way a pure liquid can be obtained free of dissolved impurities. Distillation is also used for separating mixtures of two or more liquids by exploiting differences in their boiling points. The mixture is heated until it reaches the temperature at which one of the components boils. The fraction is then collected as it distils over at a constant temperature. When this component has been removed, the temperature again begins to rise until it reaches the boiling point of another component, which is collected in a separate receiver. *See also* fractional distillation, vacuum distillation, destructive distillation.

2. Process used to separate the substances composing a mixture; it involves a change of state, as of liquid to gas, and subsequent condensation (see States of Matter). A simple distillation apparatus consists of three parts: a flask in which the mixture is heated, a condenser in which the vapor is cooled, and a vessel in which the condensed vapor, called distillate, is collected. Upon heating, the substances with a higher boiling point remain in the flask and constitute the residue. When the substance with the lowest boiling point has been removed, the temperature can be raised and the process repeated with the substance having the next lowest boiling point. The process of obtaining portions (or fractions) in this way is called fractional distillation. In destructive distillation various solid substances, such as wood, coal, and oil shale, are heated out of free contact with air, and the portions driven off are collected separately. Distillation is used in refining Petroleum and in preparing alcoholic beverages.

Distortion, 1. In general, any undesirable change in the waveform of a signal.

2. A type of aberration of lenses in which the magnification of the optical system varies with the distance from the axis. Thus a square object may have an image in which the square has sides that bulge outwards *(barrel distortion)* or curve inwards *(pincushion distortion)*. Distortion can be reduced by suitable stops.

Distributive law. In mathematics, a law governing the interaction of certain pairs of operations. The distributive law of multiplication over addition states that the equation $a \times (b + c) = (a \times b) + (a \times c)$ holds for all numbers, a, b, and c.

Disulphur dichloride. *See* sulphur chloride.

Diterpene. *See* terpene.

Diuretic. A drug that increases the rate of urine formation by the kidney, thus eliminating excess sodium and water from the body. Diuretics are given to treat such diseases as high blood pressure, glaucoma, and heart insufficiencies.

Diurnal circle. Apparent path followed by a star due to the earth's rotation on its axis. The stars appear to move from east to west on the celestial sphere in concentric circular paths centered at the celestial poles.

Divalent (bivalent). Having a valency of two.

Divergene (of a vector). The scalar product ∇, F, where ∇ is the operator del. It is usually written div F and is equal to i.e. $F/\partial x + j \partial f/\partial y + k \partial F/\partial z$.

Divergent. Designating a sequence or series that is not convergent.

Diverging lens. *See* lens.

Diving, Deep-sea. Descent into deep water for extended periods, used in a variety of commercial, scientific, and military activities. Helmeted diving suits were first devised in England in the 17th cent., and improved versions are still in use today. The scuba (acronym for Self-Contained Underwater Breathing Apparatus), invented by J.Y. Cousteau, which enables divers to carry their own air supply, allows much greater mobility. The bathyscaphe, developed by Auguste Piccard, was the first self-contained diving craft. Skin diving, a popular sport, is done with a face mask and snorkel (plastic tube) just below the surface, or with a scuba at greater depths.

Diving, springboard and platform. Sport of entering the water from a raised position, often while executing tumbles and other acrobatic maneuvers. Springboard diving is done from a flexible plank either 1 or 3 m (about 3 ft 3 in. or 9 ft 7 in.) above the water. Platform diving (or high diving) is usually done from a rigid platform projecting from a tower 10 m (32 ft 10 in.) high. Both types are scored on the basis of form, execution, and difficulty of the dive. Competitive diving, a part of most aquatic sports meets, is an Olympic event for men (since 1904) and women (since 1912).

Divisor. A quantity divided into another quantity. In $a/b = c$, b is the divisor.

D-layer. *See* ionosphere.

DMA. *See* dimethylamine.

DNA or **deoxyribonucleic acid.** Nucleic Acid found in the nuclei of Cells. It is the principal constituent of Chromosomes, the structures that transmit hereditary characteristics. The amount of DNA is constant for all typical cells of any given species of plant or animal, regardless of the size or function of that cell. Each DNA molecule is a long, two-stranded chain made up of subunits, called nucleotides, containing a sugar (deoxyribose), a phosphate group, and one of four nitrogenous bases: adenine (A), guanine (G), thymine (T), and cytosine (C). In 1953 J.D. Watson and F.H. Crick proposed that the strands, connected by hydrogen bonds between the bases, were coiled in a double helix. Adenine bonds only with thymine (A—T or T—A) and guanine only with cytosine (G—C or C—G). The complementarity of this bonding insures that DNA can be replicated, i.e., that identical copies can be made in order to transmit genetic information to the next generation.

Dodecahedron. A polyhedron with twelve faces. A *regular dodecahedron* has congruent equilateral pentagons as its faces.

Dodecanoic acid. *See* lauric acid.

Dog. Carnivorous, domesticated Mammal *(Canis familiaris)* of the family Canidae. It is distinguished from other canines, e.g., Wolves, Foxes, by its worldwide distribution in close association with humans and by the enormous genetic variability within the species. It is probable that the dog was the first animal to be domesticated over 10,000 years ago and that the earliest dogs resembled the wild Australian Dingo. For centuries dogs have been selectively bred for special purposes, notably to pursue and retrieve game, as draft animals, as guides, and as companions. There are systems for classifying and breeding dogs throughout the world, many using a variation of the British system. In the U.S. the American Kennel Club recognizes 125 of the more than 200 known breeds. The breeds are grouped into six classes: Sporting Dog, Hound, Terrier, Working Dog, Toy Dog, and Nonsporting Dog. A purebred dog is one that conforms to the standards of a certain breed and whose pedigree has been recorded for a certain period of time. Dogs of mixed origin are called mongrels. Dogs mate about every six months; their gestation period is about nine weeks, and litters vary from 2 to 10 puppies.

Dogtooth violet. Originally a name for the Old World plant *Erythronium denscansis,* now used also for several North American species of the same genus of the Lily family. The most common American species is the lilylike, yellow-flowered *E. americanum,* also called adder's tongue. Dogtooth violets are unrelated to the true Violets.

Dogwood or Cornel. Common name for some members of the family Cornaceae, trees or shrubs (genus Cornus) chiefly of north temperate and tropical mountain regions. Dogwoods have inconspicuous flowers surrounded by large, showy bracts, often mistaken for petals. The flowering dog wood *(C. florida)* and the Pacific dogwood *(c., nutallii)* are cultivated as ornamentals. Their bark, rich in tannin, is used as a quinne substitute. The bunchberry, or dwarf cornel *(C. canadensis),* is a low, herbaceous wild flower of North America. The black Gum is sometimes included in the dogwood family.

Dolomite. 1. White, gray, brown, or reddish calcium magnesium carbonate mineral [$CaMg(CO_3)_2$], commonly crystal line.

2. Carbonate rock composed chiefly of the mineral dolomite, similar to Limestone but harder and heavier. Dolomite formations are widespread and are notable in the region of the Alps called the Dolomitesis, where the rock was first studied. Dolomite is used chiefly as a building stone, for the manufacture of refractory furnace linings, and as basic magnesium carbonate for pipe coverings.

Dolphin. Aquatic Mammal, any of the small, toothed, gregarious Whales of the family Delphinidae. They include the breaked dolphin, the killer whale, the pilot whale, and 12 freshwater species in South America and 5 Asia. Fishlike in form, dolphins breathe through a dorsal blowhole. They propel themselves by means of powerful flukes, steering with a dorsal fin and navigating with the aid of echolocation. Dolphins are exceptionally friendly toward humans, and their high order of intelligence and their complex language have long been the subject of study.

Domagk, Gerhard. (domak), 1895-1964) German chemist and phathologist. Because of a Nazi decree, he had to decline the 1939 Nobel Prize in physiology or medicine, awarded for his discovery of the efficacy of prontosil, the forerunner of the Sulfa Drugs, in treating streptococcal infections. He received the Nobel Prize medal (but not the prize money) in 1947.

Dome. A roof circular or (rarely) elliptical in plan, usually hemispherical in form, placed over a square, circular, or other space. Ancient examples are found at Mycenae and in Sicily, but the Romans apparently were the first developers of the form. Their constructions culminated in the Pantheon (2nd cent. A.D.). The use of the pendentive, however—essential to placing a dome over a square — was discovered by the Byzantine builders of Hagia Sophia at Constantinople (A.D. 532-37). Islamic architects, under Byzantine influence, built many domes, e.g., the Dome of the Rock, Jerusalem. The persian or onion dome is best seen at the Taj Mahal . Roman and Byzantine influence converged in the designers of the Italian Renaissance. A circular drum was usually interposed between pendentive and dome, to give greater elevation; the dome was then topped with a lantern. Noted domes include Brunelleschi's for the Cathedral at Florence (1420-36) and Michelangelo's for St. Peter's, Rome (Completed 1590), in the U.S. C. Bulfinch's dome for the Massachusetts state capitol, Boston, established the style. Modern domes employ a wide variety of materials and designs.

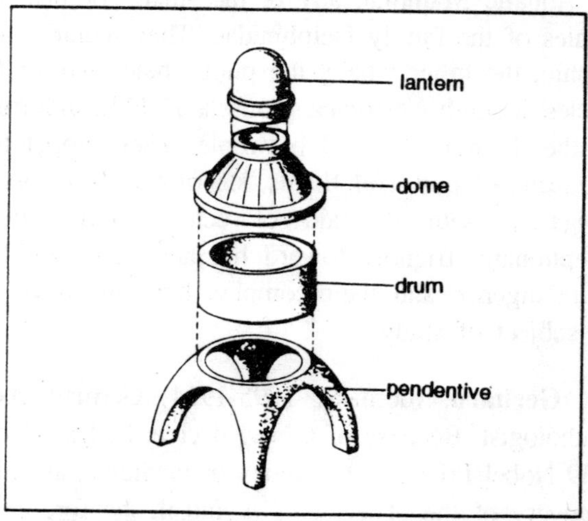

Fig. D-5. Parts of a dome.

Donor. **1.** An atom that donates electrons in an extrinsic semiconductor, thus increasing the number of electrons in the conduction band.

2. The atom or molecule that supplies the pair of electrons in forming a coordinate bond. *Compare* acceptor.

Doping. The addition of small amounts of impurity to a semiconductor to modify its electrical properties. Doping can be effected by diffusion or by ion implantation.

Doppler Effect. A change in frequency observed for a wave when relative motion occurs between the source and the observer. If the observer is moving towards the source, the frequency has a higher value: if the observer is moving away the frequency is less. The apparent frequency f is given by $f = f(c - v_o)/(c - v_s)$, where *f* is the true frequency, *c* the velocity of the waves, v_o the observer's velocity, and v_s the source's velocity. The effect occurs for all forms of wave motion, including electromagnetic waves. [After Christian Johann Doppler (1803-53), Austrian physicist.]

Doppler Effect. Change in the wavelength (and frequency) of a wave as a result of the motion of either the source or receiver of the waves. If the source and the receiver are approaching each other, the frequency of the wave will increase and the wavelength will be shortened—sounds will be higher in pitch and light will be bluer. If the source and receiver are moving apart, sounds will become lower-pitched, and light will appear redder. Astronomers analyze Doppler shifts of light and radio waves to measure the velocities and (indirectly) distances of remote objects.

Dose. A measure of exposure to ionizing radiation. The *exposure dose* is the flux of radiation to which anything is exposed. It is measured in roentgens. The *absorbed dose* measures the amount of absorption of radiation: it is the energy absorbed per unit mass. The unit is the rad. The effect of radiation on human tissue is given by the *equivalent dose*. This is the absorbed dose multiplied by a factor that expresses the biological effect of the radiation used. It is measured in rems.

Dosemeter. Any of various instruments or devices used to measure radiation dose. The techniques in common use include measurement of the blackening of a photographic film, determination of the amount of chemical reaction produced in a solution, use of ionization chambers, and measurements of induced thermoluminescence.

Dosimetry. Measurement of radiation dose using a dosemeter.

Double Refraction (birefringence). A phenomenon in which certain anisotropic crystals split a ray of incident light into two refracted rays. One ray, the *ordinary ray,* obeys the normal law of refraction. The other ray, the *extraordinary ray,* has a refractive index that depends on direction in the crystal. The two rays are both plane-polarized at right angles to each other. Double refraction occurs because the crystal is not isotropic and the light propagates at different velocities in different directions: the ordinary ray propagates with spherical wavefronts and the extraordinary ray with ellipsoidal wavefronts. Crystals that show double refraction have either one or two directions along which the light is not doubly refracted. Such a direction is an *optic axis.*

Double Salt. A mixed salt formed for crystallizing two simple salts in equivalent proportions. The alums are common examples of double salts.

Down's Syndrome. Congenital disorder characterized by moderate to severe Mental Retardation, slow physical development, and flattish skull and facial features, giving an "Oriental" appearance (hence the alternate name for the condition, Mongolism). The syndrome is caused by genetic transmittance of an extra chromosome. It can occur at any time, but the risk increases significantly with parental age, particularly if the mother is over 35. Amniocentesis can be used to detect the disorder in the fetus. Few people with Down's syndrome survive beyond age 35.

DPA. Diphenylamine.

Dobereiner's Triads. Groups of three chemically related elements in which one element has properties intermediate between those of the other two and also has a mass number that is the average of the other two. Examples of such triads are F, Cl, and Br, and Li, Na, and K. They represent an early attempt at classifying the chemical elements. [After Johann Wolfgang Dobereiner (1780-1849), German chemist].

Dragon. Mythical beast, usually represented as a huge, winged, fire-breathing reptile; prominent in the folklore of many peoples. The highest achievement of a hero in medieval legend, e.g., Saint George, was slaying a dragon. The beast is usually associated with evil; the dragon of the Book of Revelations gave rise to its use as a symbol of Satan. However, it can also be benevolent, as in ancient China.

Dragonfly. Large predatory Insect of the order odonata. They have chewing mouthparts, and four membranous, netveined wings. Dragonflies are found primarily in the tropics, but occur worldwide except in polar regions. They are strong fliers with elongated bodies that may reach 5 in. (12.7 cm) in length. They prey on mosquitoes and other insects. In the Permian period one species had a wingspan of 2½ ft (76 cm).

Dream. Mental activity .associated with the rapid-eye-movement (REM) period of sleep. It generally consists of visual images and may reflect somatic disturbances (e.g., indigestion) or external stimuli (e.g., the ringing of an alarm clock). REM sleep accounts for about half of the newborn's total sleeping time, but the proportion decreases with age to about two hours for every eight-hour period of the adult's sleep. In primitive and ancient cultures, dreams played an extensive role in myth and religion. Sigmund Freud emphasized dreams as keys to the makeup of the individual. He distinguished between the experienced content of a dream and the actual meaning of the dream, which is largely concealed from the dreamer. Jung held that dreams are not limited to the personal Unconscious but may also be shaped by innate mental structures, called archetypes, that originate in the collective unconscious of the human species.

Drfit. Deposit of clay, gravel, sand, and boulders, transported and laid down by glaciers. Some drift is stratified, or sorted by size, with coarser particles nearer the point of origin. Till, the greater part of drift, consists of unstratified heaps of rocks. Drifts can take many forms (e.g., Drumlins, Eskers, Moraines). Large sections of North America and continental Europe are covered by drift.

Drug Addiction and Drug abuse. Chronic or habitual use of any chemical substance to alter states of body or mind for other than medically warranted purposes. Among the drugs with potential for abuse are narcotics, including morphine, opium, heroin, and methadone; depressants such as Alcohole, Barbiturates, and sedative, stimulants such as Cocaine and Amphetamines; Hallucinogenic Drugs; and Marijuana. Nicotine and Caffeine can also be abused. An individual is said to be addicted if a physical dependence on a given drug develops and if withdrawal symptoms are experienced when the drug is discontinued or its dose decreased. True physical addiction is known to occur with the narcotics and depressants; psychological dependence, with or without physical symptoms, can develop with many other drugs, such as Tranquilizers. The hallucinogens can also cause traumatic experiences and trigger psychotic reactions, including paranoia. Treatment for drug addiction includes Methadone programs and participation in therapeutic communities (e.g., Synanon and Phoenix House) with other addicts who are giving up drugs. The question of what constitutes drug abuse depends on the cultural and social context. In some countries, narcotic use in the form of opium smoking is common and not considered a serious drug problem; in others, hashish or related compounds are widely used. In most industrialized nations, however, the use of many of these drugs is illegal and associated with criminal behavior.

Drugs. Substances used to cure, alleviate, diagnose, or prevent disease. Before 1900 only a few drugs were used scientifically, among them alcohol, ether, Morphine, Digitalis. Quinine, Iron, Iodine, Mercury, diphtheria antitoxin, and smallpox vaccine. Since then, and particularly since World War II, many new drugs and classes of drugs have been developed, making Chemotherapy an essential part of medical practice. Such drugs include Antibiotics. Sulfa Drugs; cardiovascular drugs, including propanolol and other Beta Blockers; Diuretics; Anticoagulants; whole blood, plasma, and blood derivatives; various smooth muscle relaxants and smooth muscle stimulants; immunosuppressive Drugs; Hormones, such as thyroxine, Insulin, and Estrogen; Cortisone and other Corticosteroids; Oral Contraceptives; Vitamins; Analgesics; poison antidotes; and various stimulants and depressants, among them Narcotics, Amphetamines, and Barbiturates. Drugs are derived from many sources, organic and inorganic. Alkaloids, hormones, vaccines, and antibiotics come from plants and animals; other drugs are synthesized in the laboratory or extracted from inorganic compounds. There are two marketing classes of drugs; ethical drugs, for which prescriptions are needed, and Patent Medicines, which are sold over the counter without prescription. The publication in the U.S. giving drug properties, actions, uses, dosages, and standards of strength and purity is the *United States Pharmacopoeia—National Formulary.* The *Physician's Desk Reference* compiles information supplied by drug companies about their products. The Food and Drug Administration regulates the testing and marketing of new drugs in the United States. The scientific study of drugs, their actions, and effects is Pharmacology.

Drum. A direct-access storage device used in computers, consisting of a cylinder coated with a layer of magnetic iron oxide. The drum rotates on a spindle and the data is stored on a number of tracks around the cylinder's surface, access being made by a set of fixed heads, one for each track.

Drypoint. Intaglio printing process in which lines are scratched directly into a metal plate with a needle; also, the print made from such a plate. The method is often used in combination with Etching. In drypoint, the burr raised by the needle produces a rich, velvety effect and allows relatively few good prints to be made. Among the greatest masters of the technique are Durer, Rembrandt, Whistler, and Picasso.

Dry Rot. Fungus disease of timber, typified by the destruction of Celluose, with discoloration and eventual crumbling of the wood. Dry not requires moisture for growth and occurs in improperly seasoned wood or where ventilation is poor or humidity high.

Drift Tube. *See* Linear Accelerator.

Duck. Wild and domestic waterfowl of the family Anatidae, which also includes the Goose and Swan. It is hunted and bred for its meat, eggs, and feathers. Strictly speaking *duck* refers to the female; the male is a *drake.* Ducks range from 16 to 22 in. (36 to 56 cm) in length and have waterproof feathers, with a thick layer of down underneath, and webbed feet. They are usually divided into three groups; surface-feeding—such as the mallard, wood duck, and teal, which frequent ponds and quiet waters; diving—such as the Canvasback and eider, found on bays, rivers, and lakes; and fish-eating, or mergansers, which also prefer open water.

Dulong and Petit's Law. The principle that the atomic heat of all solid elements is approximately equal to 3R, where *R* is the gas constant. The value is 25 J K^{-1} mol^{-1}. The law holds for elements with simple crystal structures at normal temperatures. At lower temperatures the atomic heat (molar heat capacity) is proportional to the cube of the thermodynamic temperature. [After Pierre Dulong (1785-1838) and Alexis Petit (1791-1820), French physicists.]

Dumas' Method. 1. A technique for measuring the densities of vapours by weighing the sample in a large bulb of known volume and mass.

2. An analytical technique for determining the amount of nitrogen in organic compounds. The sample is mixed with copper oxide and heated to oxidize the nitrogen present to nitrogen oxides. These are passed over hot copper, which reduces them to nitrogen gas. This estimated by a volumetric method. [After Jean Baptise Andre Dumas (1800-84), French chemist.]

Duralumin. 1. ® An alloy of aluminium and small amounts of copper (about 4%) magnesium, manganese, and silicon.

2. Alloy of Aluminum with Copper, Magnesium, and Manganese. A heat treatment causes a reaction between the aluminum and magnesium, increasing hardness and tensile strength. Because of its lightness and other properties, duralumin is widely used in the aircraft industry.

Dwarf. A dwarf star; a star of small diameter and relatively high density. It is usually some ten magnitudes fainter than an ordinary star of the same spectral type.

Dwarfism. Condition in which an animal or plant is smaller than normal size, in humans, it usually results from a combination of genetic factors and endocrine malfunction. Pituitary dwarfism is caused by a deficiency of pituitary growth hormone. Typically, the pituitary dwarf, or midget, has normal body proportions, mental capacity, and sexual development. The most famous dwarf of recent times was "General" Tom Thumb. *Cretinism,* a type of dwarfism accompanied by mental retardation and body distortion, results from an insufficiency of thyroid hormone.

Dye. 1. A colourant applied to a fabric, paper, leather, etc., from solution. Dyes are usually organic compounds owing their colour to the presence of conjugated double bonds in their molecules. In some types of dye metal atoms are also present. For the colour to be fast the substrate must have an attraction for the dye, a property known as *substantivity.* Dyes can be classified into *substantive dyes,* which are held on the substrate by ionic forces or intermolecular attraction, and *reactive* *dyes,* which actually form covalent bonds with the substrate. Other classifications are into mordant dyes, direct dyes, acid dyes, basic dyes, and vat dyes.

2. Natural or synthetic substance used to color various materials, especially textiles, leather, and food. Natural dyes are obtained from plants (e.g., indigo), from animals (e.g., cochineal), and from minerals (e.g., ocher). Used since ancient times, natural dyes have been largely replaced by synthetic dyes, first made in the 19th cent. Most of these are made from coal tar. Although some materials, e.g., silk and wool, can be colored simply by being dipped in the dye, others, including cotton, require a mordant. The process by which dyes become "attached" to the material they color is not well understood. It is thought either that a chemical reaction takes place between the dye and the fibre or that the dye is absorbed.

Dynamite. Explosive made from Nitroglycerin and various amounts of an inert, absorbent filler, e.g., sawdust. Often ammonium nitrate or sodium nitrate is added. Invented in 1866 by Alfred B. Nobel, dynamite is usually pressed in cylindrical forms and wrapped in an appropriate material, e.g., paper or plastic. The charge is set off with a detonator.

Dynamometer. Any of various instruments for measuring power. The power generated by a machine can be determined in a variety of ways including measurement of the heat produced by friction when working under a load, measurement of electrical energy produced by turning a generator, and measurement of the twist produced in a rotating shaft.

Dyne. A CGS unit of force equal to the force that, applied to a body with a mass of one gram, produces an acceleration of one centimeter per second per second. It is equivalent to 10^{-5} newton.

E

E. Irrational Number occurring widely in mathematics and science, approximately equal to the value 2.71828; it is the base of natural, or Naperian, Logarithms. Like Π (see pi), *e* is transcendental, i.e., not a Root of any algebraic equation. It is defined at the Limit of expression $(1 + {}^{1}/_{n})^{n}$ as *n* becomes infinitely large. Expressions of the form e^{x}, known as the exponential function, occur in applications ranging from statics to nuclear physics.

Eagle. Large, predatory Bird of the Hawk family, found worldwide. Similar to but larger than the Buzzards, eagle are solitary and believed to mate for life. Their beaks are nearly as long as their heads, and their wingspan may reach $7^{1}/_{2}$ ft (228 cm.) The American bald eagle (*Haliaetus leucocephalus*), an endangered species, is brown with white head, neck, and tail plumage. The golden, or mountain eagle, (genus *Aquila*), is sooty brown with tawny head and neck feathers. Unlike the bald eagle, its legs are feathered to the toes.

Ear. Organ of hearing and equilibrium. The human ear consists of outer, middle, and inner parts. The outer ear, the visible portion, includes the skin-covered flap of cartilage (auricle) and the auditory canal, which leads to the eardrum. The middle ear contains three small bones, or ossicles, known because of their shapes as the hammer, anvil, and stirrup. The inner ear contains the cochlea, which houses the sound-analyzing cells, and the vestibule, with the organs of balance. In the course of hearing, sound waves enter the auditory canal and strike the eardrum, causing it to vibrate. The waves are concentrated as they pass through the ossicles to a small opening leading to the inner ear. The vibration sets in motion fluid within the cochlea. This agitates a delicate membrane,

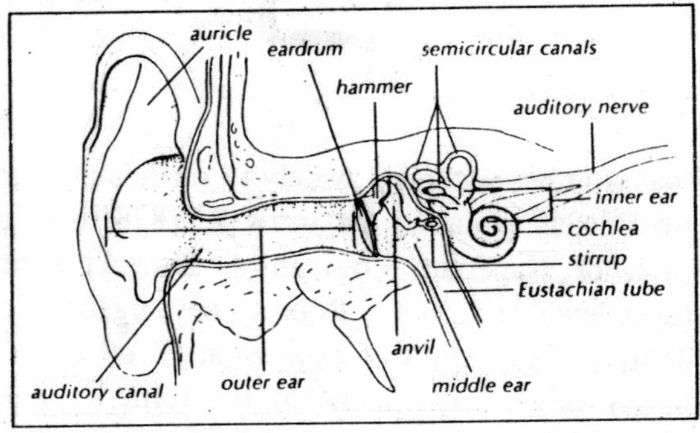

Fig. E-1. Ear.

stimulating thousands of arsensory hair cells, which in turn stimulate the auditory nerve to send impulses to the brain. Three fluid-filled semi-circular canals and two saclike organs, the utriculus and the sacculus, are the chief organs of balance and orientation; as with hearing, stimulation of sensory hair cells in these organs stimulates nerve impulses.

Earth. 1. The Third planet of the solar system in order of succession from the sun. The earth is not quite spherical, being flattened slightly at the poles because of axial rotation. It has a central core surrounded by a mantle and overlaid by a thinnish outer crust. Some 70% of the surface is covered by water and the whole planet is surrounded by a gaseous atmosphere.

Equatorial diameter: 12 756.776 km; mass: 5.98 X 10^{24} kg; relative density: 5.52; period of axial rotation: 23 hours 56 minutes 4 seconds; period of revolution round the sun: 365.256 days; mean distance from the sun: 149 598 000 km; inclination of equator: 23°27' to the ecliptic.

2. Fifth largest Planet of the Solar System and the only one known to support life. Its mean distance from the Sun is c.93

million mi (150 million km). The change of seasons is caused by the tilt (23.5°) of the earth's Equator to the plane of the orbit. The earth is surrounded by an envelope of gases, mostly oxygen and nitrogen, called the Atmosphere. Gravitational forces have molded the earth into a spherical shape that bulges slightly at the equator (equatorial diameter: c.7,926 mi/12,760 km; polar diameter: 7,900 mi/12,720 km). Studies indicate that the earth consists if concentric layers that differ in size, chemistry, and density. The outer shell, or crust, consists of the Continents and the Ocean basins. The crust is broken into vast plates that slide around on a plastic zone, or asthenosphere, within the middle shell, or mantle (*see* Continental Drift; Plate Tectonics). At the centre of the earth is an outer core, believed to be liquid, and an inner, solid core. The earth is estimated to be 4.55 billion years old, and its origin continues to be a controversial subject. The earth has one natural satellite, the Moon.

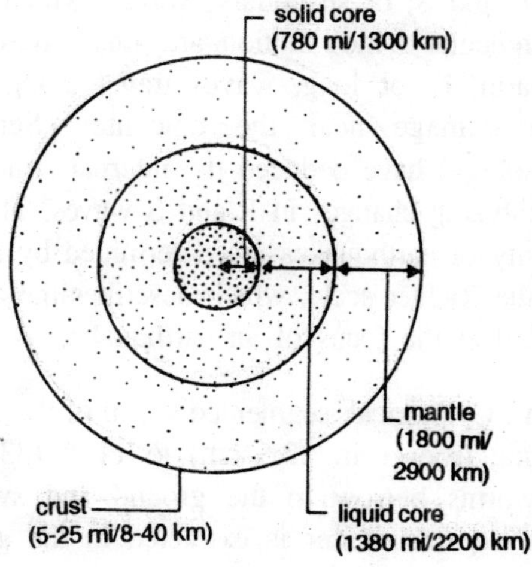

Fig. E-2. Cross section of the earth, showing its shells.

Earth. An electrical connection between a piece of equipment and the earth. Such connections are made for safety reasons so that an appliance cannot become "live" if a fault develops or they are made to produce a fixed reference potential, considered to be zero. In the U.S. the term *ground* is used.

Earthshine. Light from the sun reflected from the earth's surface producing a faint illumination of the dark part of the moon.

Earthquake. Trembling or shaking movement of the earth's surface. Great earthquakes usually begin with slight tremors, rapidly increase to one or more violent shocks, and diminish gradually. The immediate cause of most shallow earthquakes is the sudden release of stress along a Fault, or fracture, in the earth's crust, resulting in the movement of opposing blocks of rock past one another. This causes vibrations to pass through and around the earth in wave form (see Tsunami). The subterranean origin of an earthquakes is its focus; the point on the surface directly above the focus is the epicenter. Waves generated by earthquakes are of three types. Both P, or primary, waves, which are compressional and are the fastest, and S, or secondary, waves, which cause vibrations perpendicular to their motion, are body waves that pass through the earth. L, or long, waves travel along the surface and cause damage near the epicenter. Seismologists (see Seismology) have deduced the internal structure of the earth by analyzing changes in P and S waves. The magnitude and intensity of earthquakes are determined by the use of scales, e.g.; the Richter scale, which describes the amount of energy released at the focus of an earthquake.

Earthworm. Cylindrical, segmented worm of the class Oligochaeta. Ranging from 1 in. (2.5 cm) to 11 ft (330 cm) in length, earthworms burrow in the ground and swallow soil from which organic matter is extracted in the gizzard. They are important to agriculture in aerating and mixing the soil.

Ebonite. *See* vulcanite.

Ebony. Common name for the Ebenaceae, a family of trees and shrubs widely distributed in warm climates. The genus

Diospyros includes the ebony and persimmon trees. Ebony wood, valued since ancient times, is dark and hard; it is used extensively in cabinetmaking. Some species (e.g., *D. hirsuta*) have wood striped with black or shades of brown (called calamander wood or variegated ebony). The persimmons bear edible fruit that when unripe is astringent and when ripe is soft and pulpy, making it difficult to market. Persimmon wood has a limited use in the manufacture of objects (e.g., golf club heads) requiring hardwood.

Ebullition. The bubbling or effervescence of a liquid; boiling.

Eccentricity. *See* conic.

Echelon. A Type of diffraction grating made by stacking glass plates so that there is a constant offset between each plate to form a series of equal steps with widths of about 1 mm. The plates have to be of equal thickness to within less man one wavelength of light. The echelon can be used as a reflection grating or light can be transmitted through it parallel to the steps. The device has a very high resolution and is used for studies of the fine structure of spectral lines.

Echidna or Spiny Anteater. Primitive egg-laying Mammal, or Monotreme, of New Guinea, E, Australia, and Tasmania. Covered with sharp quills, the grayish-brown echidna protects itself by rolling into a ball. It may reach 18 in. (46 cm) in length. A rapid burrower, the echidna probes for ants and termites with its sensitive muzzle and long, sticky tongue.

Echinoderm. Marine invertebrate animal of the phylum Echinodermata, having external skeletons of calcareous plates just under the skin, no head, and a unique water-vascular (ambulacral) system with tube-feet. Echinoderms are radially symmetrical and lack specialized excretory organs and reproduce sexually. Echinoderms have extensive powers of regeneration of lost or injured parts. All members of the phylum live on the sea floor. The phylum includes the Starfish,

Brittle Stars, Sea Urchins and Sand Dollars, Sea Cucumbers, and Crinoids.

Echo. 1. A sound repeated as a result of the reflection of sound waves from a surface.

2. The reflected signal in radar.

Echolocation. A technique, such as radar or sonar, in which an object is located by the echo reflected from it.

Echo sounding. A technique for measuring the depth of the sea by transmitting a pulse of sound and measuring the delay before the reflected pulse is detected.

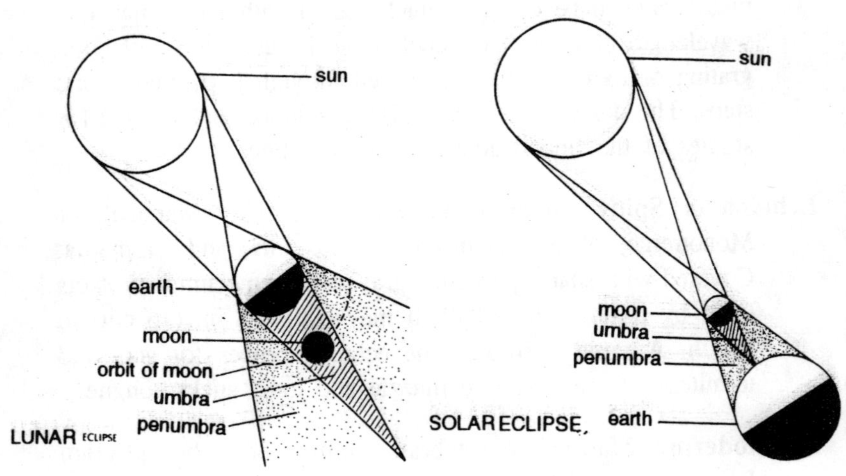

Fig. E-3. Eclipses.

Eclipse. The temporary blocking of the light from the sun (*solar Eclipse*) or moon (*lunar eclipse*) as viewed from the earth. Solar eclipses occur when the earth, moon, and sun are directly in line so that the dark body of the moon passes in front of the sun and casts its shadow on the earth's surface. Observers directly in the moon's shadow (*umbra*) see a total eclipse: those on the fringes of it (*penumbra*) see only a partial one. A lunar eclipse takes place when the moon passes into the shadow cast by the earth. Again the eclipse may be total or partial depending on how much of the moon is obscured.

Eclipse. Partial or total obscuring of one celestial body by the shadow of another. A *lunar eclipse* occurs when the earth blocks the sun's light from the moon; a *solar eclipse* occurs when the moon blocks the sun's light from some area on the earth. Because the earth and moon shine only by the reflected light of the sun, each casts a shadow into space in the direction away from the sun. The shadow consists of a cone-shaped area of complete darkness, called the umbra, and a larger area of partial darkness that surrounds the umbra, called the penumbra. A lunar eclipse is said to be total or partial depending on whether the moon passes completely into the umbra of the earth's shadow or remains partly in the penumbra. A solar eclipse is said to be total or partial depending on whether the observer is in the umbra or the penumbra of the moon's shadow. Total solar eclipses have particular scientific importance, providing information about the motions of the moon and about the surface phenomena of the sun and permitting tests of Albert Einstein's general theory of Relativity. An annular solar eclipse occurs when the moon is so distant from the earth that the umbra of its shadow is too short to reach the earth's surface; the sun is thus seen as a bright ring completely surrounding the moon's dark disk. Solar and lunar eclipses are possible only when the moon crosses the earth's orbital plane at a time when the sun, moon, and earth are aligned (at new and full moon; see Syzygy).

Eclipsed conformation. *See* conformation.

Ecliptic. The great circle on the celestial sphere that marks the earth's orbit around the sun and along which the sun apparently moves relative to the constellations of the zodiac. The circle of the ecliptic is inclined to the plane of the celestial equator at an angle of 23°27'. Because the orbits of the planets in the solar system are roughly in the same plane these objects also tend to lie in the region of the ecliptic.

Ecliptic. The great circle on the celestial sphere (see Astronomical Coordinate Systems) that lies in the plane of the earth's Orbit. Because of the earth's yearly revolution around the sun, the sun appears to move in an annual journey along the ecliptic. The obliquity of the ecliptic is the inclination (about 23-1/2°) of the plane of the ecliptic to the plane of the celestial equator. The constellations of the Zodiac are those through which the ecliptic passes.

Ecology. Study of the interrelationships of organisms and the physical environment. Within the Biosphere, the basic unit of study is the ecosystem—a community of plants and animals in an environment supplying the raw materials for life, i.e., chemical elements (or food) and water. An ecosystem is delimited by climate, altitude, latitude, water, and soil characteristics, and other physical conditions. The energy for fueling life activities reaches the earth in the form of sunlight. By Photosynthesis green plants capture that light energy and store it in the chemical bonds of carbohydrates, fats, and proteins. Some of the energy is acquired by plant-eating animals, and a fraction of it is passed on to predatory animals. Such sequences, called food chains, overlap at many points. Once spent, the energy for life cannot be replenished except by further exposure of green plants to sunlight. The chemicals of the life are continually recycled by such processes as photosynthesis, Respiration, and nitrogen fixation (see Nitrogen Cycle). Disruption of these cycles by natural causes such as drought or by Population can disturb the balance of an entire ecosystem. An ecosystem that has reached a stable and self-perpetuating stage is known as a *climax community*. When extensive and well-defined, the climax community is called a *biome*. Examples are Tundra, Savanna (grassland), Dessert, and forests. Stability is attained through a process known as succession, whereby a relatively simple community, such as lichen- and algae-covered rocks, gives way over time to one more complex, such as a forest or tundra.

Eddy Currents. Alternating electric currents induced in the cores of transformers and other equipment by the changing magnetic fields associated with the equipment. The current causes a power loss in the device. Eddy currents can be reduced by using laminated iron, made up of varnished strips, or by using a ferrite core.

Eddington, Sir Arthur Stanley. 1882-1944, English astronomer and physicist. He was Plumian professor of astronomy (from 1913) and director of the observatory (from 1914) at Cambridge Univ. Eddington made major contributions to the study of the evolution, motion, and internal constitution of stars, and was one of the first physicists to grasp the theory of relativity, of which he became a leading exponent. A prolific writer on science, he wrote such works as Mathematical Theory of Relativity (1923) and The Internal Constitution of the Stars (1926).

Edelweiss. Perennial plant (genus *Leontopodium*) of the composite family, found at high altitudes in the mountains of Europe, Asia, and South America. It has woolly-white floral leaves and small yellow disk flowers surrounded by silvery bracts. It is esteemed in Europe as a symbol of purity, and in Switzerland is protected by law.

Edison, Thomas Alva. 1847-1931. American inventor; b. Milan, Ohio. Edison was a genius in the practical application of scientific principles and one of the most productive inventors of his time—despite only three months of formal schooling and an increasing deafness throughout most of his life. Among his most important inventions were the carbon Microphone (1877), the Record Player (patented 1878), and the Kinetoscope (see Motion Pictures). HIs most significant contributions, however, were his development of the first commercially practical incandescent lamp (8179) and his design for a complete electrical distribution system for Lighting and power, culminating in the installation (1881-82) of the world's first central electric-light power plant in New York City. His New Jersey workshops (at Menlo Park and West Orange) were forerunners of the modern industrial research laboratory, in which teams of workers, rather than a lone inventor, systematically investigate a problem.

Edta (ethylenediaminetetracetic acid). A white powder, $(HOOCCH_2)_2N\text{-}(CH_2)_2N(CH_2COOH)_2$, used as a chelating agent.

Eel. Any of a large order (Anguilliformes) of Fish with a long, snakelike body, elongate dorsal and anal fins, no pelvic fins, and scales either lacking or embedded in the skin. The male American eel reaches 2 ft (61 cm) in length; the female, 4 ft (122 cm). Most, including the morays and conger eels, are marine; one family, Anguillidae, is freshwater. All eels spawn in the sea, the eggs hatching into transparent, ribbonlike larvae that drift about feeding until they metamorphose into small eels, or elvers; freshwater species typically undertake long migrations to spawn in the sea. Freshwater eels are an important food item in some parts of the world.

Effective value. *See* root mean square.

Effervescence. Emission of bubbles of gas from a liquid, especially when the gas is formed *in situ* by a chemical reaction.

Efficiency. *See* machine.

Efflorescence. The production of a powdery solid: either from crystals, by the spontaneous loss of water of crystallization, or from a solution by evaporation. Sometimes the term is used to refer to the powdery deposit itself rather than the process.

Effort. *See* machine.

Effusion. The flow of a gas from a vessel through a small orifice, with no pressure gradient, due to thermal motion of the gas molecules. The volume of gas molecules. The volume of gas effusing in unit time is inversely proportional to its molecular weight. Thus effusion can achieve partial separation of gases of different molecular weights.

Egret. Name for several Heron species. Egrets were nearly exterminated by hunters seeking their white, silky plumage (which develops during the mating season) for millinery. The American egret has straight plumes, c.21 in. (52.5 cm) long, on its back. The snowy egret, or snowy heron (*Leucophoyx thula*), the most beautiful and most hunted, has curved plumes on the head, breast, and back.

EHF. Extremely High Frequency.

Ehrlich, Paul. 1854-1915, German bacteriologist. For his work in immunology he shared with Elie Metchnikoff the 1908 Nobel Prize in physiology or medicine. He discovered salvarsan for the treatment of syphilis and made valuable contributions in hematology, cellular pathology, cancer study, and the use of dyes in microscopy and treatment of disease.

Eiffel Tower. Structure designed by the French engineer Alexandre Gustave Eiffel (1832-1923) for the Paris Exposition of 1889. The tower is 984 ft (300 m) high and consists of an open iron framework on four masonry piers; these piers support four columns that unite to form the shaft.

Eigenfunction. One of a set of allowed wave functions of a particle in a given system as determined by wave mechanics.

Eigenvalue. One of a set of allowed energies of a particle in a given system as determined by wave mechanics.

Einstein. A unit of energy of electromagnetic radiation equal, for radiation of a particular frequency, to the energy of one mole of photons. The value depends on the frequency (ν) of the radiation and is equal to $Nh\nu$, where N is the Avogadro constant and h is the Planck constant.

Einstein, Albert. 1879-1955, American theoretical physicist; b. Germany; recognized as one of the greatest physicists of all time. He became (1914) titular professor of physics and director of theoretical physics at the Kaiser Wilhelm Institute in Berlin. The Nazi government confiscated (1934) his property and revoked his German citizenship because he was Jewish, and in 1940 Einstein became an American citizen, holding a post at the Institute for Advanced Study in Princeton from 1933 until 1945. Although an ardent pacifist, he urged Pres. Franklin Roosevelt to investigate the possible use of atomic energy is bombs. In one of three important 1905 papers, he explained Brownian Movement on the basis of his study of the motion of atoms. His special theory of Relativity (1905) dealt with systems or observers in uniform (unaccelerated) motion with respect to one another. He asserted (1911) the equality of Gravitation and Inertia and formulated (c. 1916) a general theory of relativity and included gravitation as a determiner of the curvature of a Space-time continuum. Einstein contributed to the development of Quantum Theory, postulating (1905) light quanta (Photons), on which he based his explanation of the Photoelectric Effect, and developing and quantum theory of Specific Heat. Working on a unified field theory, he attempted to explain gravitation and electromagnetism with one set of laws. For his work in theoretical physics, notably on the photoelectric effect, he received the 1921 Nobel Prize in physics.

Einsteinium. Symbol: Es. A transuranic actinide element made by neutron bombardment of plutonium. The most stable isotope, ^{254}Es, has a half-life of about 320 days.

Einsteinium (Es). Radioactive element, discovered in 1952 by A. Ghiorso and colleagues in residue from a thermonuclear explosion. A member of Actinide Series, weighable amounts have since been prepared by neutron bombardment of plutonium.

Einstein photoelectric equation. *See* photoelectric effect.

Einstein's equation. The equation $E = mc^2$, giving the energy E equivalent to mass m. where c is the velocity of light. The interconversion of matter and energy occurs in nuclear fission and fusion and in pair production and annhiliation.

Einstein shift. A slight displacement of lines in a star's spectrum towards longer wavelength, caused by an intense gravitational field. The Einstein shift in distinct from a red shift caused by the Doppler effect. It is predicted by the general theory of relativity.

Einthoven galvanometer. A type of galvanometer consisting of a thin conducting filament stretched so that it passes midway between the poles of a magnet. A current passed through the filament caused a deflection, which is measured by a high-power microscope let through one of the pole pieces.

Elastance. Th reciprocal of the electrical capacitance. Elastance is measured in drafts.

Elastic. 1. *See* elasticity.

2. Denoting a collision between bodies in which the total kinetic energy of bodies is conserved and no energy is converted into internal energies of the bodies. An elastic collision between a particle and an atom occurs when the particle is scattered without causing excitation or ionization. Collisions that do produce energy changes of this type are said to be *inelastic*.

Elasticity. The property of substances that return to their original size and shape when a deforming force is removed. The force causing the change is measured by the stress and the deformation it produces is measured by the strain. If the material recovers its dimensions completely, i.e. the strain disappears when the stress is removed, it is said to be *elastic.* Above the *elastic limit* (*see* Hooke's Law), permanent deformation is produced and the material is *inelastic.* A substance that can be deformed easily with no recovery when the strain is removed is said to be *plastic.* The elastic properties of a material are measured by its modulus of elasticity.

Elastomer. An elastic material, such as a natural or synthetic rubber or some similar synthetic resin.

E-layer. *See* ionospher.

Electret. A solid dielectric that exhibits permanent dielectric polarization. Such materials maintain their polarization when the electric field is removed.

Electrical image. A hypothetical point change or system of changes used in the solution of electrostatic problems. If charge $+q$ is placed a certain distance from a conductor, the force between the charge and the surface is the same as that which would occur between the charge and its "image" in the surface—a charge of-q placed as far behind the surface plane as the original charge is in front of it.

Electric and Magnetic units. Unit used to express the magnitudes of various quantities in electricity and magnetism. Three systems of such units, all based on the Metric System, are commonly used. One of these the mksa-practical system, is defined in terms of the units of the mks system and has the Ampere of electric current as its basic unit. The unit of the system—the Volt, Ohm, Watt, and farad—are those commonly used by scientists and engineers to make practical measurements. The two other systems, now being gradually abandoned, are both based on the cgs system. Electrostatic units (cgs-esu) are defined in a way that simplifies the description of interactions between static electric charges; there are no corresponding magnetic units in this system. Electromagnetic unit (cgs-emu), on the other hand, are defined especially for the description of phenomena associated with moving electric charges, i.e., electric currents and magnetic poles.

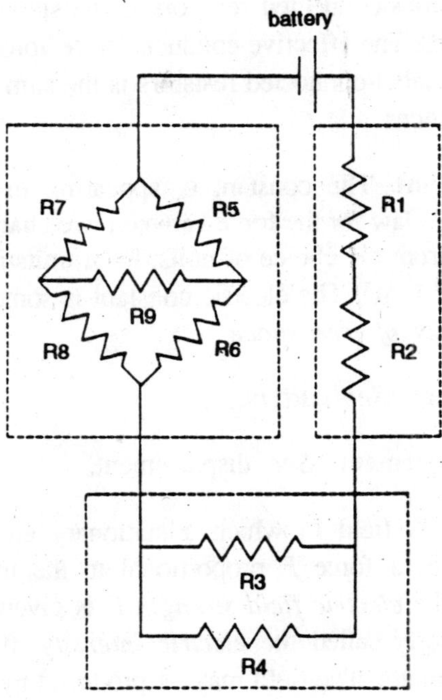

Fig. E-4. Electric circuit: The resistance R_1 and R_2 are connected in series; the same current flows through both. The resistance R_3 and R_4 are connected in parallel; the same voltage appears across both. The resistance TR_5, R_6, R_7, R_8, and R_9 are connected in a configuration that is neither series nor parallel. The three two-terminal subcircuits enclosed in broken lines are all connected in series with the battery.

Electric circuit. Unbroken path along which an electric current may flow. A simple circuit consists of a voltage source, such as a battery (see Cell, in electricity) or a Generator, whose terminals are connected to those of a circuit element, such as Resistor, through which current can flow. More complex circuits include additional sources or elements and perhaps Switches, so interconnected that, when appropriate switches are closed, each element is included in a closed path that also contains a source. Series, parallel, and non-series-parallel connections are illustrated in the figure. The effective Resistance of two series-connected resistors is the sum of the individual resistances. The effective conductance reciprocal of resistance) of two parallel-connected resistors is the sum of the individual conductances.

Electric constant. The constant ε_0 appearing in the equation of Coulomb's law for the force between two charges in a vacuum. It arises from the choice of units: in SI units its value is 8.854 16 x 10^{-12} F m^{-1}. The electric constant is sometimes called the *permittivity of free space.*

Electric current. *See* current.

Electric displacement. *See* displacement.

Electric field. A field in which a stationary electric charge (q) experiences a force F proportional to the magnitude of the charge. The *electric field strength* E is given by $E = F/q$. E was formerly called the *electric intensity*. It is measured in volts per metre. The field may be produced by other stationary electric charges, in which case it is an *electrostatic field.* The force on a stationary charge as a result of the electric field from another stationary charge i.e. given by Coulomb's law. Electric fields can also be created by a changing magnetic field. Thus, if a magnetic field is changed in the neighbourhood of a conductor it induces an electromotive force.

Electric fish. Any of many unrelated species of Fish that have electric organs, usually made of modified muscle tissue, that are capable of producing from 450 to 600 volts of electricity. This electricity is used to detect and paralyze prey, repel enemies, navigate, and possibly communicate. Important electric, navigate, and possibly communicate. Important electric fish include the electric Eel, the electric Catfish, and the electric Ray, or torpedo.

Electricity. Class of phenomena arising from the existence of Charge. According to modern theory, most Elementary Particle of matter possess charge, either positive or negative. Two particles of like charge, both positive or both negative, repel each other; two particles of unlike charge are attracted (see Coulomb's Law). The electric Force between two charged particles is much greater than the gravitational force between the particles. Many of the bulk properties of matter are ultimately due to the electric forces among the particles of which the substance is composed. Materials differ in their ability to allow charge to flow through them. Those that allow charge to pass easily are conductors (see Conduction), whereas those that allow extremely little charge to pass through are called insulators (see insulation), or Dielectric. A third class of materials, called Semi-conductors, is intermediate. Electrostatics is the study of charges, or charged bodies at rest. When positive or negative charge builds up in fixed positions on objects, certain phenomena can be observed that are collectively referred to as static electricity. The charge can be built by rubbing certain objects together, such as silk and glass or rubber and fur; the friction between these objects causes Electrons to transfer from one to another with the result that the object losing electrons acquires a positive charge and the object gaining electrons acquires a negative charge. Electrodynamics is the study of charges in motion. A flow of electric charge constitutes an electric current. In order for a current to exist in a conductor, there must be an Electromotive Force (emf), or Potential difference, between the conductor's ends. An electric Cell, A Photovoltaic Cell, and a Generator are all sources of emf. An emf source with an external conductor connected from one of the source's two terminals to the other constitutes an Electric Circuit. Direct current (DC) is a flow of current in one direction at a constant rate Alternating current (AC) is a current flow that increases in magnitude from zero to a maximum, decreases back to zero, increases to a maximum in the opposite direction, decreases to zero, and then repeats this process periodically.

The number of repetitions of the cycle occurring each second is defined as the frequency, which is expressed in Hertz (Hz). The frequency of ordinary household current in the U.S. in 60 cycles per sec (60 Hz) and electric device must be designed to operate at this frequency. In a solid, the current consists not of a few electrons moving rapidly but of many electrons moving slowly; although this drift of electrons is slow, the impulse that causes it moves through the circuit, when the circuit is completed, at nearly the speed of light. The movement of electrons in a current is not steady; each electron moves in a series of stops and starts. In a direct current, the electrons are spread evenly through the conductor; in a alternating current, the electrons tend to congregate along the conductor surface. In liquids, gases, and semiconductors, current carries may be positively or negatively charged.

Electric motor. A device for obtaining mechanical energy from electrical energy by electromagnetic induction. Alternating-current motors are similar to generators, with the current supplied to the armature through slip rings. Motors of this type rotate at the speed that would generate current of the same frequency as the current supplied. Such device (*synchronous motors*) run at a speed proportional to the frequency of the power supply.

Direct-current motor are similar to d.c. generator, with current supplied through a segmented commutator. The coil generating the magnetic field are supplied with the same current. In a *series-wound* motor the coils are connected in series with the armature, giving a motor that loses speed as its load increases. A *shunt-wound* motor has its field coils connected in parallel with the armature. Shunt-would motors do not lose speed at increased loads.

Most large a.c. motor are *induction motors,* which have a stator supplied with alternating current in such a way as to produce an effective rotating magnetic field. One method of doing this is by mounting three coils at 120° and using a three-phase alternating-current supply. The rotor is mounted in the centre of the stator and typically consists of a steel bar with copper bars parallel to the axis connected at each end by copper rings (called a *squirrel cage*). The magnetic field induces a current in the rotor and interaction of this field and the current exerts a torque on the rotor. Induction motors are nonsynchronous.

Electrocardiography. Science of recording and interpreting the electrical activity of the Heart by means of a device called a cardiograph. Electrodes (leads) attached to the exterminities and the left chest pick up electrical currents from heart muscle contractions. The currents are registered in wave patterns on a light-sensitive-film recording known as an electrocardiogram, cardiogram, or simply EKG or ECG. Deviations in the normal height, form, or duration of the wave pattern indicate specific disorders; thus the EKG is an important aid in diagnosing many heart diseases. The first practical device for recording the activity of the heart was developed by W. Einthoven in 1903, for which he won (1923) the Nobel Prize in physiology or medicine.

Electrochemical equivalent. The mass of a substance that is deposited or liberated from a solution of its ions by passage one coulomb of electricity.

Electrochemistry. The branch of chemistry concerned with the study of electrolysis, electrolytic cell, and similar applications or properties of ions in solution.

Electrochemistry. Science dealing with the relationship between electricity and chemical changes. Of principal interests are the reactions that take place between Electrodes and the Electrolytes in electric and electrolytic cells (see Electolysis) and that take place in an electrolyte as electricity passes through it.

Electrode. A conducting plate, wire grid, etc., used to emit, collect, or control the flow of charged particle in a liquid, gas, or semiconductor. The positive electrode of a system is the *anode:* the negative electrode is the *cathode.* The charge carriers may by electrons, holes, or ions. Electrodes are used in transistors, electrolytic cells, thermionic valves, gas-discharge tubes, and similar devices.

Electrode. Terminal, usually in the form of a wire, rod, or plate, through which electric current passes between metallic and nonmetallic parts of an Electric Circuit. The electrode through which current passes from the metallic to the nonmetallic conductor is called the anode; that through which current passes from the nonmetallic to the metallic conductor is called the cathode. An electrode may be made of a metal e.g., copper, lead, platinum, silver, or zinc, or of a nonmetal, commonly carbon.

Electrodeposition. The deposition of a dissolved substance on an electrode by electrolysis, particularly the plating of a cathode by neutralization of metal ions. The cathode process may be written $Ag^+ + e = Ag$.

Electrode potential. The electric potential developed on a metal electrode when it is in equilibrium with a solution of its ions. For example, if a metal rod is placed in a solution of metal ions two competing processes occur. The atoms can enter solution as ions, leaving electrons in the electrode. Conversely, ions in solution can gain electrons from the metal and be deposited as atoms. At equilibrium these process occur at equal rates and the potential on the metal rod depends on the excess or deficiency of electrons. In fact, the potential of such a half cell cannot be measured absolutely and electrode potential are usually measured against a hydrogen electrode, which is assigned an electrode potential of zero. Under standard conditions the value is called the *standard electrode potential.*

Electrodynamometer. *See* dynamometer.

Electroencephalography. Science of recording and analyzing the electrical activity of Brain. The recording known as an electroencephalogram, of EEG, is made through electrodes by an electroencephalograph and appears as characteristic wave patterns. An important aid in diagnosing brain disorders and malfunctions, it has contributed to understanding normal brain function, sleep, and the effect of Drugs and such activities as Biofeedback. The electrical activity of the brain was first demonstrated by the German psychiatrist Hans Berger in 1929.

Electroforming. A technique for producing thin intricate metal articles by electrodeposition of a layer of metal onto a shaped former. The former is first coated with a conducting layer of carbon and made the cathode in an electrolytic cell.

Electroluminescence. Fluorescence caused by bombardment with electrons. Electroluminescence is the process occurring in neon lights and other types of fluorescent lamp.

Electrolysis. Chemical change produced by passing an electric current through a conducting solution or fused ionic substance. Such liquids (electrolytes) contain ions of opposite charge. When a voltage is applied across two electrodes in the electrolyte, positive ions drift towards the cathode and negative ions towards the anode, thus carrying an electric current. At the electrode, positive ions can gain electrons and negative ions can lose electrons, forming neutral atoms or radicals. Alternatively, atoms of the electrode material can ionize and pass into solution in the electrolyte.

Electrolysis. Passage of an electric current through a conducting solution or molten salt (either is a type of Electrolyte) that is decomposed in the process. When a cathode, or negative electrode, and an anode, or positive electrode, are dipped into a solution, and a direct-current source is connected to the electrodes, the positive ions migrate to the negative electrode and the negative ions migrate to the positive electrode. At the negative electrode each positive ion gains an electron and becomes neutral; at the positive electrode each negative ion gives up an electron and becomes neutral. The migration of ions through the electrolyte constitutes the electric current flowing from one electrode to the other. Electrolysis is used in the commercial preparation of various substances, e.g., chlorine by the electrolysis of a solution of common salt, and hydrogen by the electrolysis of water. The electrolysis of metal salts is used for plating. See also Electrochemistry.

Electrolyte. A liquid that conducts electricity. Electrolytes may either be solutions or melten solids. In both cases the current is carried by ions drifting under the influence of an electric field produced between the electrodes. Electrolytes are of two types. *Strong electrolytes* have a high conductivity. They are made by dissolving electrovalent compounds, such as sodium chloride, in water or by melting ionic crystals. They are simply mixtures of the ions present in the solid. *Weak electrolytes* are produced by dissolving covalent compounds that partially dissociate in water to give ions. Examples of this type are weak acids and bases, such as acetic acid and ammonia.

Electrolyte. Electrical conductor in which current is carried by Ions rather than free electrons (as in a metal). Electrolytes include water solutions of acids, bases, or Salts; certain pure liquids; and molten salts.

Electrolytic. Pertaining to electrolysis or eletrolytes.

Electrolytic gas (detonating gas). An explosive mixture of hydrogen and oxygen in the proportions 2:1 by volume. It can be produced by electrolysis of water.

Electrolytic separation. A method of separating isotopes by using electrolysis. The rate at which ions are discharged at an electrode depends on the ion's mass. The technique is used for separating hydrogen from deuterium by eletrolysing water. At the cathode, the reaction $H^+ + e = H$ is slightly faster than $D^+ + e = D$. Consequently, over a long period of time the water become enriched with heavy water.

Electromagnet. A magnet made by winding a large coil of wire around a core of soft iron, so that a magnetic field is produced when a current is passed through the coil.

Electromagnet. Device in which an electric current, passing through a wire coil wrapped around a soft iron core, produces a magnetic field. The magnetic-field strength produced depends on the number of turns of the coil of wire, the size of the current, and the magnetic permeability of the core. Electromagnets lose their magnetism when the current is discontinued. See also Superconductivity.

Electromagnetic induction. The production of an electric current in a conductor by a changing magnetic field. The phenomenon was originally observed by Faraday when a magnet was moved in and out of a coil of wire connected to a galvanometer. It is found that an electromotive force is induced when a conductor cuts a magnetic flux. This can happen in a variety of ways. For example, a circuit placed in a changing magnetic field has an induced e.m.f. as the magnetic flux linked with the circuit changes. This is the origin of the current in the secondary coil of a transformer. A similar effect is produced by moving a magnet towards a coil of wire. Electromagnetic induction also occurs hen a conductor moves through a magnetic field, thus cutting the lines of force. This is the mechanism producing current in a generator. The magnitude of the induced e.m.f. is proportional to the rate of cutting of magnetic flux.

In any circuit in which there is a changing electric current there is an associated changing magnetic field and this in turn may induce an e.m.f. in the circuit. The influence of such a field on the circuit in which the current is flowing is called *self-induction.* The flux ϕ linked with the circuit is proportional to the current flowing: i.e. * = LI, where L is a constant depending on the geometry of the circuit. When the current is changing the induced e.m.f. is given by $E = -LdI/dt$, dI/dt being the rate of change of current. The minus sign appears because the e.m.f. opposes the current flow (in accordance with Lenz's law). L is the *self-inductance* of the circuit. Similarly, a current flowing in one circuit can induce an e.m.f. in a neighbouring one. This is *mutual induction.* The induced e.m.f. is given by $E_2 = -M\ dI_1/dt$. Here dI_1/dt is the rate of change of current in the first circuit, E_2 the e.m.f. induced in the second circuit, and M is the *mutual inductance* of the circuits. *See also* induction.

Electromagnetic interaction. A type of interaction between elementary particles arising as a result of their electric charge. Stationary charges interact by an electrostatic field and moving charges produce a magnetic field. The electromagnetic interaction is about 200 times weaker than the strong interaction. It operates at all distances and is inversely proportional to the square of distance. The effect is thought to be caused by the exchange of virtual photons (see virtual particle).

Electromagnetic pump. A device for pumping liquid metals through pipes by passing an electric current across the diameter of the pipe and applying a strong magnetic field at right angles to the direction of current flow. The liquid metal then experiences an induced force along the pipe, causing it to move. Electromagnetic pumps are used for moving the liquid sodium coolant in nuclear reactors. They are simple and have no moving parts.

Electromagnetic radiation. Waves of energy (*electromagnetic waves*) consisting of electric and magnetic fields vibrating at right angles to the direction of propagation of the waves. Electormagnetic waves are produced by accelerating charged particles and their properties are described by Maxwell's equations. They have a velocity, c, equal to $2.997\ 925 \times 10^8$

metres per second and their properties depend on their frequency (or wavelength). The whole frequency range of elctromagnetic radiation is called the *electromagnetic spectrum,* which ranges from radio waves to gamma rays and includes visible radiation (light). An alternative, and complementary, model of electromagnetic radiation is of streams of photons.

Electomagnetic radiation. Energy radiated in the form of a Wave comprised of an electric field interacting with a magnetic field. Electromagnetic radiation is the result of the acceleration of a charged particle. It does not require a material medium,

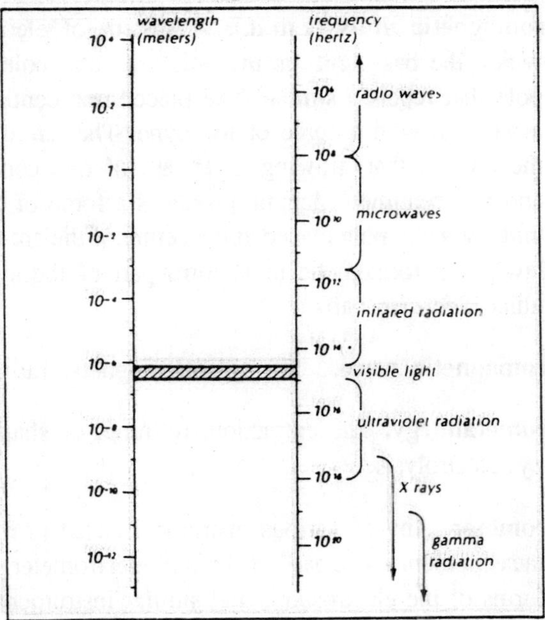

Fig. E-5. Types of electromagnetic radiation.

and can travel through a vacuum. The theory of electromagnetic radiation was developed by James Clerk Maxwell and published in 1865, although his ideas were not accepted until Heinrich Hertz proved the existence of radio waves in 1887. In order of decreasing wavelength and increasing frequency, the various types of electromagnetic radiation are Radio waves, Microwaves, Infrared Radiation, visible Light, Ultraviolet Radiation, X rays, and Gamma Radiations. The possible sources of electromagnetic radiation are directly related to wavelength; long radio waves are produced by large antennas such as those used by broadcasting stations; much shorter visible light waves are produced by the motions of charges within atoms; the shortest waves, those of gamma radiation, result from changes within the nucleus of the atom. The individual quantum of electromagnetic radiation is known as the Photon.

Electromagnetic units (e.m.u.). A system of electrical units in which the base unit is the unit magnetic pole. defined as a pole that repels a similar pole placed one centimeter away in a vacuum with a force of the dyne. The e.m.u. of current is the current that, flowing in an arc of one centimeter radius and one centimetre length, produces a force of one dyne on a unit magnetic pole placed at the centre of the arc (*see* Ampere's law). Electromagnetic units form part of the system of CGS units. *See also* ab-.

Electromagnetic wave. *See* electromagnetic radiation.

Electrometallurgy. The extraction, refining, or shaping of metals by electrolysis.

Electrometer. Any of various instruments used to make electrical measurements. The earliest types of electrometer were modified forms of the electroscope and similar instruments that depend for their action on electrostatic attraction or repulsion. Examples are the electrostatic voltmeter and the quadrant electrometer. Most of these are now obsolete except for specialized uses and modern electrometers are electronic devices for amplifying small voltages. The important factor in such instruments is a high input impedance, so that no current is drawn from the circuit investigated. This was formerly achieved by using valves (as in the valve voltmeter) or by using a vibrating-reed electrometer. Now, special field-effect transistors are used. Besides the determination of low voltages, electrometers can be used for small currents (10^{-7} amperes) by measuring the voltage drop they produce across a very high standard resistance. Similarly, they can be used for measuring very high resistances (up to 10^{18} ohms).

Electromotive force (e.m.f). Symbol. *E*. The energy supplied by a source of current in driving unit charge around an electrical circuit. It is measured in volts. The electromotive force of a current supply is not the same as the potential difference across the terminals when current is flowing because of the internal resistance of the source.

Electromotive series. A series of the metals arranged in decreasing order of their tendency to form positive ions by a reaction of the type $M = M^+ + e$. The energy of this process is the ionization potential for an isolated metal atom. Usually the elements are arranged in order of their electrode potential, which gives the tendency to form ions in aqueous solution. The series for some of the more common elements is Li, K. Ba, Ca, Na, Mg, Al, Zn, Cr, Fe, Sn, Pb, H, Ti, Cu, Ag, Hg, Au. The more electropositive the element, the higher its position in the series Metals that have high positions can displace lower metals from their salts. For example, iron displaces metallic copper from copper sulphate. $Fe + Cu^{2+} + Cu$. Similarly, metals such as zinc that are higher than hydrogen in the series will dissolve in dilute acids with evolution of hydrogen: $Zn + 2HCl = ZnCl_2 + H_2$. Metals such as copper will not displace hydrogen. The series is sometimes called the *electropositive* or *electrochemical series*.

Electron. An elementary particle with a negative charge equal to that of the proton and a rest mass of 9.1095 X 10^{-31} kg (about 1/1836 that of the proton). Electrons are present orbiting the nuclei of all atoms and are also produced by beta decay. They are produced in electric discharges and in thermionic emission.

Electron. Elementary Praticle carrying a unit charge of negative electricity. An Atom consists of a small, dense, positively charged nucleus surrounded by electrons that whirl about it in orbits, forming a cloud of charge. Ordinarily there are just enough negative electrons to balance the positive charge of the nucleus, and the atom is neutral. If electrons are added or removed, a net charge results, and the atom is said to be ionized (see ION). Atomic electrons are responsible for the chemical properties of matter (*see* Valence). The electron was discovered in 1897 by Joseph John Thomson, who showed that cathode rays are composed of electrons. The electron is the lightest known particle having a non-zero rest mass. The positron, the electron's antiparticle (see Antimatter), was discovered in 1932.

Electron affinity. Symbol: *A*. The energy produced by attachment of an electron to an atom or molecule to form a negative ion. The electron affinity of an atom is positive if energy is released. It is usually measured in electronvolts.

Electron deficient. Denoting a chemical compound in which there are too few bonding electrons to satisfy the simple rules of valency. The boron hydrides. for example have molecules in which two boron atoms are linked by an intermediate hydrogen atom with only one pair of electrons shared between the three atoms. This is an example of a multicentre bond.

Electron diffraction. Diffraction of electrons by atoms in molecules or crystals. The original example was the Davisson-Germer experiment in which electrons were diffracted by the surface atoms of a metal. This technique is used for studying solid surfaces. Individual molecules in the gas phase can also produce diffraction patterns from their atoms and these can be used for measuring bond lengths and bond angles of molecules.

Electronegative. Tending to form negative ions by gain of electrons.

Electronegativity. Symbol: X. The tendency of an atom in a molecule to attract electrons in chemical bonds. The quantity is defined in two ways. In one it is equal to $(1 + A)/2$, where I is the ionization potential and A is the electron affinity. The other definition is based on bond energies between two elements Y and Z. If E_{YZ} is the bond energy of a Y-Z bond and E_{YY} and E_{ZZ} are the energies of Y-Y and Z-Z bonds, the difference in electronegativity of Y and Z is given by $X_Y - X_Z = E_{YZ}$ — $E_{ZZ}E_{YY}$. Values are stated with reference to fluorine, which is given the value 4.

Electronegativity. In chemistry, tendency for an atom to attract a pair of electrons that it shares with another atom. If the pair of electrons is not shared equally, i.e., if they spend more time with one atom than with the other atom, the favored atom is said to be more electronegative. Nonmetals have much higher electronegativities than metals; of the nonmetals, fluorine is the most electronegative, followed by oxygen, nitrogen, and chlorine.

Electron gun. A device for generating a beam of electrons for use in television tubes, cathode-ray tubes, klystrons, mass spectrometers, etc. The source of the device is usually a wire filament which is electrically heated and which emits electrons by thermionic emission into a surrounding vacuum. The electrons are formed into a narrow beam by arrays of focusing and accelerating electrodes.

Wlectronic. 1. Relating to electronics. **2.** Relating to electrons. Thus the electronic energy levels of a molecule are those defined by the orbital energy of the electrons.

Electronic Game. Device that challenges eye-hand coordination in a setting designed to provide entertainment and, sometimes, education. Made possible by the development of the Microprocessor, electronic games are marketed in various sizes, such as hand-held one-player models, home-video types consisting of cartridges that are inserted in modules attached to television sets, and freestanding arcade versions. Most of their appeal comes from the Computer Program that synchronizes flashing lights and a variety of rhythmic auditory stimuli with the action portrayed on a graphic display (see Computer Graphics). The games may be contested among several players, or an individual may engage in a test of skill against the Computer. Game subjects include sports (e.g., baseball and football); action warfare, often space-oriented; and such classics as contract bridge, chess, and poker.

Electronic music. Term applied to compositions whose sounds are either produced or modified electronically. The early examples of electronic compositions, called concrete music, were taped montages of electronically altered sound obtained by microphone from nonelectronic sources, such as voices and street noise. After the perfection of the tape recorder in the 1940s, music was composed with tones that were electronically produced as well as electronically manipulated. In 1955 the synthesizer was developed, a single unit with numerous generating and modifying capacities that enable composers to generate new sounds and to combine these sounds with great precision. Contemporary composers of electronic music include Karlheinz Stockhausen, Luciano Berio, and Burno Maderna. Since the 1950s, composers such as John Cage, Milton Babbitt, and others have also experimented with Computer-generated music. Many rock musicians play electronic instruments, i.e., the instrument is connected to an amplifier and its sound is fed through speakers (see Rock Music).

Electronics. A branch of electrical engineering; the science and technology based on and concerned with the controlled flow of electrons or other carriers of electric charge, especially in electron Tubes and Semiconductor devices. The miniaturization (see Microelectronics) and savings in power brought about by the invention (1948) of the Transistor and the subsequent development of Integrated Circuits have allowed Electric Circuits to be packaged more densely, making possible compact Computers, advanced Radar and navigation systems, more reliable Radio and Television receivers, advanced sound-reproducing systems, and electronic automotive accessories.

Electron Lens. An arrangement of coils or electrodes producing a magnetic or electric field used to focus a beam of electrons.

Electron microscope. A device in which a magnified image of a sample is produced by using beams of high-energy electrons rather than light. Above a certain magnification, images produced by optical microscopes lack detail because of the limit of resolution. This depends on the wavelength of the light. Electron microscopes have higher resolution because at energies in the keV range the electrons have much shorter wavelengths than light (*see* de Broglie equation(. Magnifications as high as 200 000 can be achieved and objects as small as 0.2-0.5 nm can be resolved. Two main types of electron microscope are used.

In the *transmission electron microscope* a narrow beam of electrons is focused on the sample in a vacuum by a combination of magnetic and electrical lenses. The sample is thin and electrons passing through it are focused onto a fluorescent screen, thus producing the image.

In the *scanning electron microscope* a thick sample can be used. The primary beam of electrons is deflected by a varying field so that the sample is scanned by the beam and secondary electrons ejected from its surface are focused onto the screen (*see* secondary emission). Although the scanning microscope has a lower resolution than the transmission instrument it does give a three-dimensional image of the object.

Electron microscope. Instrument that produces a highly magnified image of a small object, using an accelerated electron beam held within a vacuum to illuminate the object, and electromagnetic fields to magnify and focus the image. (By contrast, the optical Microscope uses visual light for illumination, and glass lenses for magnifying and focusing.) First developed in Germany c. 1932, the transmission electron microscope exploits the very short wavelengths of accelerated electrons, which are thousands of times shorter than those of light, to magnify an object up to one million times without losing definition. Images produced by the microscope are thrown onto a fluorescent screen or are photographed. The more recently developed scanning electron microscope constructs an image on a Cathode-Ray Tube as the surface of a specimen is gradually illunimated by a television scanning probe.

Electron multiplier. An electronic device for detecting electrons by using secondary emission. It consists of an evacuated tube containing a number of electrodes, each held at a higher positive voltage than the one preceding it. An electron hitting the first electrode ejects two or more secondary electrons. These are accelerated to the second electrode where they produce more electrons, and so on. The number of electrons is multiplied by the chain of electrodes (called *dynodes).* Some types of electron multiplier consist of a narrow spiral glass tube with an insulating coating inside. A very high potential difference is applied between the ends of the coating and electrons hitting one end of the tube are accelerated down it, making successive collisions with the walls and producing more and more secondary electrons. *See also* photomultiplier.

Electron optics. The study of the focusing and deflection of beams of electrons in cathode-ray tubes, electron microscopes, and similar devices.

Electron-probe microanalysis. A technique for the analysis of small quantities of a solid sample by bombarding the specimen with a fine beam of electrons and monitoring the characteristic X-rays produced. The elements present can be determined and the method can be quantitative, especially for elements of high atomic number. Quantities as small as 10^{-16} kg can be detected with a beam of about 1 micrometre diameter. The technique is particularly useful for investigating the surface structure of solids.

Electron spin resonance (ESR). A technique similar to nuclear magnetic resonance, for investigating the spins of the electrons in atoms. The radiation absorbed lies in the microwave region.

Electron tube. An electronic component in which a current or voltage is controlled by flow of electrons or ions between electrodes in a vacuum or gas. Electron tubes are usually thermionic valves or types of discharge tube.

Electron tube. Device consisting of a sealed enclosure in which electrons flow between Electrodes separated either by vacuum (in a vacuum tube) or by an ionized gas at low pressure (in a gas tube). The two principal electrodes of an electron tube are called the anode and cathode. The simplest vacuum tube, the Diode, contains only these two electrodes. When the cathode is heated, it emits a cloud of electrons, which are attracted to the positive polarity of the anode and constitute the current through the tube. Because the anode is not capable of emitting electrons, no current can flow in the reverse direction, and the diode acts as a Rectifier.. In the vacuum triode, small signals applied to a third electrode, called a grid, placed between the cathode and anode cause large fluctuations in the current between the cathode and anode. A triode can thus act as a signal Amplifier. Although formerly the key elements of Electric Circuits, electron tubes have been almost entirely displaced by Semiconductor devices. *See* also Cathode-Ray Tube.

Electronvolt. Symbol: eV. A unit of energy equal to the energy acquired by an electron when it is accelerated through a potential difference of one volt. It is used in atomic and nuclear physics, particularly in experimental studies of charged particles.

Electrophilic. Having or involving an affinity for negative electric charge. An electrophilic reagent (or *electrophile)* is one that acts by attacking a negative region of a molecule. Examples are halogen molecules and the nitronium ion (NO_2^+), which occurs in nitrations. Electrophilic reactions are ones involving such a reagent. Thus, in an *electrophilic substitution* a hydrogen atom is displaced by an electrophile: this occurs in the substitution of benzene and its compounds. In *electrophilic addition* the molecule is initially attacked by an electrophile: this happens in the addition of halogens to alkenes. *Compare* nucleophilic.

Electrophoresis (cataphoresis). The motion of charged colloidal particles through a stationary fluid under the influence of an applied electric field. Electrophoresis is used as a method of separating or analysing mixtures, particularly mixtures of proteins. Typically, two electrodes are placed in contact with a sheet of paper moistened with salt solution. A small amount of sample is applied to the paper and different components of the mixture migrate at different rates.

Electrophorus. A simple device for generating an electric charge by friction and electrostatic induction. It consists of a flat insulating plate used in conjunction with a flat metal plate with an insulating handle. The insulating plate is given an electrostatic charge by friction and the metal plate is placed on top of it. Charge on the insulator polarizes the metal so that it has one charge on its lower surface and an equal and opposite charge on its upper surface. The electric charge on the upper surface is then removed by momentarily connecting it to earth: this leaves a net charge on the metal plate.

Electroplating. The process of coating a metal object with a thin layer of the same or, more usually, another metal by electrodeposition.

Electropositive. Tending to form positive ions by loss of electrons.

Electropositive series. *See* electromotive series.

Electroscope. A simple instrument for detecting an electric charge, consisting of a metal rod with two thin leaves of metal suspended from it passing vertically through an insulating stopper into a glass container. A charge applied to a plate on top of the rod causes the metal leaves to swing apart as a result of the mutual repulsion of their like charges. The leaves are usually made of gold leaf because of its extreme thinness. *See also* electrometer.

Electrostatic. Designating or involving stationary electric charges or effects produced by such charges. An *electrostatic field,* for example, is an electric field produced by a stationary charge. Electrostatic effects are those caused by forces between stationary charges and are distinguished from effects such as electromagnetism, electrolysis, and the heating effect of a current, all of which are caused by moving charges. Electricity produced by electrostatic charge is known an *static electricity.*

Electrostatic field. *See* electric field.

Electrostatic generator. A device, such as a Wimshurst machine or Van de Graaff generator, for generating an electrostatic charge.

Electrostatic induction. The separation of charge in a neutral body as a result of the effect of an electric field. If a positively charged body is brought close to a neutral insulating material it attracts it (a plastic comb for example, charged by friction, will pick up small pieces of paper). The electric field produced by the positive charge polarizes the insulator (*see* polarization) and the net effect is that the surface region nearer the positive charge has a negative charge whereas the further part of the insulator has a positive charge. The force of attraction occurs between the original positively charged body and the induced negative charge on the insulator. Electric charges can also be induced in metal objects; in this case the electrons can flow freely through the metal so one half of the object has a negative charge and the other half a positive charge (the overall charge is zero). *See also* induction.

Electrostatic precipitation. The removal of small solid particles suspended in a gas by electrostatic charging and subsequent precipitation onto a collector in a strong electric field. The process is used in cleaning air in tunnels and effluent gases in industrial chimneys.

Electrostatics. The study of electrostatic phenomena. This branch of the study of electricity is also known as *static electricity.*

Electrostatic units. (e.s.u.). A system of electrical units in which the base unit is the e.s.u. of charge, the statcoulomb, which is defined as the charge that repels an identical charge with a force of one dyne when placed one centimetre away in vacuum (*see* Coulomb's law). Electrostatic units form part of the system of CGS units. *See also* stat-.

Electrostatic voltmeter. A type of electrometer used for measuring potential differences in the kilovolt range. A metal vane suspended on a torsion wire is free to move between fixed plates, the angle of deflection being proportional to the square of the potential difference between these and the metal vane. *See also quadrant* electrometer.

Electrostriction. A change in the dimensions of a crystal of insulating material when it is subjected to an electric filed. *Compare* piezoelectric effect.

Electrovalent bond (ionic bond, polar bond). A type of chemical bond in which atoms or groups of atoms are held together by electrostatic forces between ions. A simple compound between two elements can be electrostatic forces between ions. A simple compound between two elements can be formed by transfer of electrons. For example, calcium atoms have the electron configuration 2, 8, 8, 2 and oxygen atoms have the configuration 2, 6. The calcium atom can transfer its two outer electrons to the oxygen and form a calcium ion with a double positive charge, Ca^{2+}. This has the same number of electrons as the inert gas argon. The oxygen atom gains two electrons to form a doubly charged negative ion. O^{2}. It also has the electron configuration, and thus the stability, of an inert gas (in this case neon). The ions are then held together by electrostatic forces.

Ionic bonding of this type is not confined to simple ions. Ammonium sulphate, $(NH_4)_2SO_4$, contains ammonium ions, $[NH_4]^+$, and sulphate ions, $[SO_4]^2$. Within each ion the bonds are covalent. In many ionic transition-metal compounds the positive ions do not have inert-gas configurations.

No distinct molecules are present in electrovalent compounds. A crystal of calcium oxide is simply a three-dimensional regular array of ions. Because the forces between ions are strong, ionic solids usually have high melting points. They tend to dissolve in water and other polar solvents.

Electrum. An alloy of gold and silver (15-45%) used in jewellery.

Element. A substance consisting of atoms that all have the same atomic number. Elements cannot be broken down into simpler substances. At present, over 104 elements are known although technecium (atomic number 43) and the transuranic elements are made artifically by transmutation. Of the elements in the earth's crust, oxygen (47% by weight) and silicon (28%) are the two most common. Aluminium is the commonest metal and iron, calcium, sodium, potassium, and magnesium are also fairly common. Together these eight elements make up about 99% of the earth's crust. In the universe as a whole, hydrogen is the commonest (about 90% by weight) followed by helium (about 9%).

Element. In chemistry, substance composed of Atoms all having the same number of Protons in their nuclei. This number, called the Atomic Number defines the element and establishes its place in the Periodic Table. Each element is assigned a symbol of one or two letters (see table). The total number of protons and Neutrons in the nucleus of an atom is called the Mass Number. Although all atoms of an element have the same number of protons in their nuclei, they may not all have the same number of neutrons. Atoms of an element with the same mass number make up an Isotope of the element. All elements have isotopes; over 1,000 isotopes of the elements are known. As of 1982, 108 elements were known. Only 92 elements occur naturally on earth; the others are produced artificially (see Synthetic Elements; Transuranium Elements). The chemical properties of an element are due to the distribution of electrons around the nucleus, particularly the outer, or Valence, electrons (the ones involved in chemical reactions). Chemical reaction does not affect the nucleus and thus does not change the atomic number. For this reason an element is often defined as a substance that cannot be decomposed into simpler substances by chemical means.

Elementary particles. Tiny bits of matter assumed to be the most basic constituents of the universe. Certain elementary particles combine to form an Atom, which is the basic unit of any chemical Element and from which all forms of matter are built. The first elementary particle to be discovered was the electron, identified in 1897 by Joseph John Thomson The nucleus of ordinary hydrogen was subsequently recognized as a single particle and was named the proton. The third basic particle in an atom, the Neutron, was discovered in 1932. Although models of the atom consisting of just these three particles are sufficient to account for all forms of chemical behaviour of matter, Quantum Theory predicted the existence of additional elementary particles. A search for the positron, or antiparticle (see Antimatter) of the electron, led to its detection in 1932, but a search for a particle predicted by Yukawa Hideki in 1935 led to the unexpected discovery of the mu meson, or muon, the following year. Yukawa's particle was finally discovered in 1947 and named the pi meson, or pion. Both the muon and the pion were first observed in Cosmic Rays. As the list of particles and antiparticles grew, through further study of cosmic rays and the study of the results of particle collisions produced by Particle Accelerators, four basic categories of particles were distinguished, according to their behaviour with regard to the four fundamental forces of nature; gravitational, electromagnetic, strong nuclear, and weak nuclear. A given particle experiences certain of these forces but may be immune to others. The gravitational force is experienced by all particles. The electromagnetic force is experienced only by charged particles, although it is transmitted by the Photon, which has no charge. The weak and strong nuclear forces exist only at the atomic level. Of the four classes of particles, the smallest is that of the massless bosons, which include the photon, eight types of gluons, and the hypothetical graviton. The lepton class includes twelve particles: the electron, the positron, the positive and negative muons, the tauon and its antiparticle, and the neutrino or antineutrino associated with each of these particles. The bosons and the leptons are not strongly interacting. Members of the meson class are more massive than the leptons. The mesons are the "glue" that holds nuclei together. By far the largest class of particles is the baryon class, the lightest members of which are the proton and neutron; the heavier members are the hyperons. Baryons and mesons, both strongly interacting, are sometimes considered together as hadrons. A theory in dependently proposed in 1964 by Murray Gell-Mann and George Zweig explains the properties of all known hadrons according to the assumption that hadrons are built up of other, still more fundamental particles called Quarks.

Elementary particles. Any of a number of particles of subatomic size considered as fundamental units of matter. The proton, electron, neutrino, and photon are the only stable particles. The neutron is stable when in the atomic nucleus. Many other unstable particles and resonances are known from experiments with cosmic rays and accelerators. All decay spontaneously into other particles. elementary particles are characterized by their rest mass and by other properties such as charge, spin, isospin, hypercharge, parity, and strangeness. In addition to the gravitational and electromagnetic interactions, they partake in strong and weak interactions. Those taking part in strong interactions are classified as hadrons and those taking part in weak interactions as leptons. Hadrons are further classified into baryons (including nucleons and hyperons) and mesons. Another classification, based on the spin of the particles, divides them into fermions and bosons.

Masses of elementary particles are often expressed in units of energy/c^2 (Einstein's equation). 1. MeV/c^2 = 178 X 10^{-30} kg.

Elephant. Largest living land Mammal, of Africa and Asia. It has a massive head, thick legs, and tough, gray skin. The upper lip extends into a long, flexible trunk used for drinking, spraying, conveying food to the mouth, and touching; the large ears provide an extensive cooling surface. African bull elephants may measure 13 ft (4 m) high at the shoulder and weigh 6 to 8 tons (5400 to 7200 kg.) Hunted for their great, curved ivory tusks, they have become endangered. Highly intelligent, they are used in foresting and trained as circus and zoo performers. See also Mammoth, Mastodon.

Elephantiasis. Abnormal enlargement of any part of the body due to obstruction of the lymphatic channels in the affected area. In tropical countries, the most common cause is a disease called filariasis, a result of the infestation of small roundworms, *Filaria,* in the Lymphatic System.

Elevation of boiling point. An increase in the boiling point of a solvent when a substance is dissolved in it. The elevation is a colligative property; for dilute solutions it is proportional to the number of molecules or ions dissolved. Measurement of the elevation is a method of determining molecular weights.

Elevator. Mechanical lift device for transporting people or goods from one level to another. The American inventor Elisha G. Otis demonstrated (1853) a safety device designed to prevent the fall of all elevator if its supporting cable should break; in 1861 he patented steam-powered elevator. Steam was gradually replaced, first in the early 1870s by hydraulic power (see Hydraulic Machinery) and then toward the end of the 19th cent. by electricity. Safety devices were improved and automatic controls introduced as elevator speeds increased.

Elinver. A type of steel containing chromium and nickel. Its elasticity is not affected by temperature changes and it is used for making the hair-springs of watches.

Elk. Any of several members of the Deer family. It most properly designates the largest member of the family, *Alces alces,* found in N North Aamerica, where it is called Moose, and Eurasia. The name *elk* is used in North America to designate a different animal, the Wapiti.

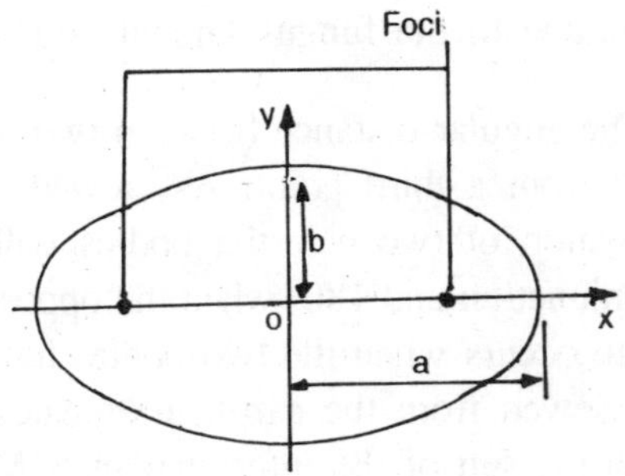

Fig. E-6. Ellipse.

Ellipse. A conic with an eccentricity less than 1. The ellipse is the locus of a point that moves so that the sum of its distances from two fixed points (the *foci)* is always constant. It has two axes of symmetry, the major and the minor axis. In Cartesian coordinates, the equation is $(x - n)^2 / a^2 + (y$ —

$m)^2/b^2$ = 1, where *(n,m)* is the centre of the ellipse, 2*a* the length of its major axis, and 2*b* the length of its minor axis. The area is P *ab*.

Ellipsoid. A surface or solid all of whose plane sections are ellipses or circles. An ellipsoid generated by rotating an ellipse about one of its axes is a *spheroid.* Rotation about the major axis produces a *prolate* spheroid while rotation about the minor axis gives an *oblate* spheroid.

Elliptical. Having the shape of an ellipse or an ellipsoid.

Elliptical galaxy. *See* galaxy.

Elliptically polarized light. *See* polarization.

Elm. Tree or shrub of the family Ulmaceae, found chiefly in the Northern Hemisphere. Elm trees (genus *Ulmus)* are tall and graceful, with fan-shaped crowns of finely subdividing branches and twigs. The American, or white, elm (*U. americana)* and the English, or Wych, elm *(U campestris)* of Eurasia are widely planted as ornamental and shade trees. Both species are vulnerable to the fungus known as Dutch elm disease.

Elongation. The angular distance between two points in the sky as measured from a third point. As viewed from a third body, the elongation of two celestial bodies will be 0° when they are in conjunction and 180° when the opposition (see Syzygy); quadrature occurs when the two bodies have an elongation of 90°. As viewed from the earth, the greatest elongations with respect to the sun of the inferior planets Mercury and Venus are 28° and 47°; the moon and the superior planets can range from 0° to 180°.

Elutriation. Purification or separation by washing, decanting, and settling. An example of elutriation is the method of panning for gold in river water.

Emanation. The radioactive gas given off by radioactive decay of radium, actinium, and thorium. The term is obsolete; the gases are now known to be radon, actinon, and thoron respectively.

Embalming. Preservation of the body after death by artificial chemical means. It was highly developed in dynastic Egypt, where immersion in a soda solution and filling of body cavities with resins and spices were common. Modern embalming grew from 17th-cent. attempts to preserve anatomical specimens. Formaldehyde, infused to replace the blood, is the most common embalming agent.

Embryo. Name for the developing young of an animal or plant. Embryology, the scientific study of embryonic development, deals with the period from fertilization until hatching or birth of an animal or, in plants, germination. In animals, early divisions produce a hollow ball one cell thick (a blastula), which later becomes a two-layered cuplike gastrula. In higher animals, a third (middle) layer of cells develops from one or both of the first two layers. The outer layer (ectoderm) gives rise to skin, scales, feathers hair, nails, and the nervous system; the innermost layer (endoderm) forms the digestive glands and lining of the alimentary tract and lungs; the middle layer (mesoderm) develops the skeletal, muscular, and connective tissue and circulatory, excretory, and reproductive systems.

Emerald. Highly valued Gem, green variety of the mineral Beryl. The finest emeralds are found in Colombia; other sources are the USSR, Zimbabwe, and Australia. The Oriental emerald is the transparent green variety of Corundum.

Emission spectrum. A spectrum produced by emission of electromagnetic radiation by matter. In forming an emission spectrum, energy has first to be supplied to the sample to produce atoms or molecules in excited states. This can be donc by heat, as in a flame; by bombardment with particles, as in a discharge tube, electric arc, or X-ray tube; or by irradiation with other electromagnetic radiation. The emitted radiation is dispersed using a prism or grating. Each line in the emission spectrum is produced by decay of excited atoms or molecules to some lower energy level. *See also* absorption spectrum, spectral series.

Emissivity. Symbol: ε. A measure of the ability of a body to radiate heat, given by the ratio of the power radiated by the body per unit area to the power radiated per unit area by a black body at the same temperature. *See* Kirchoff's law.

Emitter. The electrode connected to the central region in a junction transistor.

Emphysema. Abnormal enlargement or distension of the air sacs of the lungs, causing difficulty in breathing. Usually chronic and progressive, the condition is associated with heredity, smoking, and long-standing respiratory ailments such as chronic bronchitis. Management is aimed at increasing lung capacity and preventing and treating infection, a common complication.

Empirical. Based on the results of observation rather than on theory.

Emu or **emeu.** Large flightless Australian Bird *(Dromiceius novaehollandiae)* related to the Cassowary and Ostrich. A swift runner, it is 5 to 6 ft (150 to 180 cm) tall. Its brownish plumage is coarse and hairlike.

Emulsion. A colloidal dispersion of one liquid in another. Oil and water form emulsions; the system may be small globules of oil in a continuous water phase (an oil-in-water emulsion) or small drops of water in oil (a water-in-oil emulsion). Emulsions can be produced by agitating the components together. Soaps, detergents, and similar *emulsifiers* are often added to promote the formation of emulsions and to stabilize them.

Enamel. A siliceous substance that fuses with metal. Transparent or opaque, clear or coloured, it is usually employed to add a decorative surface. Enamel was used in making jewelry in ancient Egypt, Greece, and Rome, and was perfected in the Byzantine world, often in the Cloisonne technique. Fine enamel-work was created in 12th-cent. France and Spain. The most famous enamelist was the 16th-cent. French artist Leonard Limousin. In England from the 17th cent. on, enamel was used for miniature portraits. Enamel work declined in the 19th cent. but was revived in the 1960s.

Enantiomer. *See* optical isomerism.

Enantiomorphism. *See* optical isomerism.

Enantiotropy. The existence of a solid in two crystalline forms (*see* dimorphism) with a definite transition temperature between them. White tin for example is stable above 13.2°C; grey tin is stable below 13.2°C.

Encephalitis. General term used to describe an inflammation of the brain and spinal cord, characterized by headache, lethargy, and convulsions. Rabies and herpes viruses, among others, are common causes. For many forms of encephalitis there is no treatment, although herpes encephalitis responds to the drug adenine arabinoside.

Endangered species. Any species of animal or plant whose ability to survive is seriously in question. Human activities can contribute to such endangerment: for example, by purposeful extermination to protect livestock; unrestricted hunting to obtain hides, feathers, or food; or the use of Pesticides to protect crops. Humans have contributed to the destruction of entire habitats through, e.g., Strip mining, oil spills, water Pollution, and the draining of swamps and leveling of forests for industrial and residential development. Such destruction can endanger the lives and breeding grounds of a large number of species simultaneously. In recent years environmentalists have pressed for the establishment of new wildlife Refuges and for land use planning. In the U.S. the Endangered Species

Acts of 1966, 1969, and 1973 prohibit any trade in endangered species or their products and require that federal agencies assess the impact on wildlife of proposed projects.

Endocrine system. Body control system composed of a group of glands that keeps the internal environment stable by producing chemical regulatory substances called Hormones. In humans the major endocrine organs are the Pituitary Gland, Thyroid Gland, Parathyroid Glands, Adrenal Glands, Thymus Gland, Pineal Body, Pancreas, ovaries, and testes (see Reproductive System); the Kidneys are also sometimes included. The endocrine, or ductless, glands secrete the hormones they produce directly into the internal environment; transmitted via the bloodstream or by diffusion, the endocrine hormones act at distant points in the body. Some, such as Thyroxine from the thyroid gland, effect nearly all body cells; others, such as Progesterone from the female ovary, which regulates the uterine lining, affect only a single organ. The pituitary, sometimes called the master gland, secretes hormones that regulate many of the other endocrine glands. The regulation of body function depends on the existence of specific receptor cells in target organs — cells that respond in specialized ways to minute hormonal stimuli. The hormones act by regulating cell metabolism: By accelerating, slowing, or maintaining enzyme activity in receptor cells, they control growth and development, metabolic rate, sexual rhythms, and reproduction. The quantities of hormones are maintained by feedback mechanisms that depend on interactions between the endocrine glands, blood levels of hormones, and activities of the target organs.

Endoergic. *See* endothermic.

Endorphin. Any of a group of proteins produced by the pituitary gland, affecting mood, perception of pain, memory retention, and learning. Chemically similar to opium-derived Narcotics, endorphins were searched for and found in the 1970s after the discovery that Morphine works by attaching itself to specific receptor sites in the brain. Endorphins also attach to these receptors and appear to be the brain's own natural "painkillers." It is thought that Acupuncture reduces pain, at least in part, by stimulating the body to produce endorphins.

Endothermic. Denoting a process or chemical reaction that takes place with the absorption of heat. Nuclear processes that absorb heat are said to be *endoergic*. *Compare* exothermic.

End point. The point in a titration at which the reaction is complete, so that neither reagent is in excess. It is usually observed using an indicator.

Energy. Symbol; *E*. The capacity of a system to perform work—i.e., to move a force along its line of action. Energy is divided into potential energy, due to position, and kinetic energy, due to motion. Several other forms of energy are distinguished. *Heat* is simply the combined kinetic energies of the atoms and molecules forming the system. *Radiant energy* is energy carried by electromagnetic wave. *Electrical energy* is energy produced by a cell or generator. It is the potential energy of electrons in an electric field. *Chemical energy* is the energy stored in a compound as the kinetic and potential energies of electrons in the chemical bonds. The *law of conservation of energy* states that the total energy of a system is unchanged in any series of processes — energy is converted from one form to another. The principle has been modified by the theory of relativity in which matter is regarded as a form of energy. *See* mass-energy.

Energy. In physics, the ability or capacity to do work. Forms of energy include Heat, chemical energy, and, according to the theory of Relativity, Mass (see Nuclear Energy); other forms of energy are associated with the transmission of Light, Sound, and electricity. Energy and work are measured in the same units: joules, ergs, electron-volts, calories, foot-pounds, or some other, depending on the system of measurement being used. When a force acts on a body, the work performed (and the energy expended) is the product of the force and the distance over which it is exerted. Potential energy is the capacity for doing work that a body possesses because of its position or condition. For example, a weight lifted to a certain height has potential energy because of its position in earth's gravitational field. Kinetic energy, the energy a body possesses because it is in motion, is equal to 1/2 mv^2, where m is its mass and v is its velocity. The average kinetic energy of the atoms or molecules of a body is measured by the Temperature of the body. Energy (or its equivalent in mass) can be neither created nor destroyed, but it can be changed from one form into another.

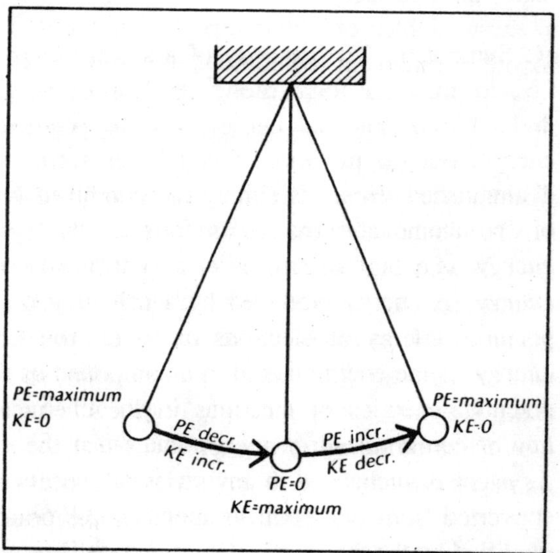

Fig. E-7. Relations between potential energy (PE) and kinetic energy (KE) for a swinging pendulum.

Energy band. A band of allowed energies for an electron in a solid. *See* band theory.

Energy level. One of a number of discrete energies that may be possessed by a nucleus, atom, molecule, or other system. for example, an electron in a hydrogen atom can only occupy certain orbits around the nucleus, each of a different energy (*see* Bohr atom). The atom can thus have a number of states: one in which the electron is in the inner orbit, another with the electron in the second orbit, etc. A specific energy level exists for each state of the atom. The state with lowest energy is the *ground state.* The higher states are *excited states.* Similarly, according to quantum theory, the vibrations between the atoms in a molecule can only occur so that the molecule has discrete vibrational energy levels.

Energy, sources of. In contemporary usage, energy is whatever can be efficiently converted into heat or motion to provide power to run machines and vehicles and to supply heat and light. Energy sources are of two basic types, renewable and nonrenewable. Most of the industrial world is presently powered by nonrenewable fossil fuels—Coal, Petroleum, and Natural Gas—that, once used, cannot be replaced. Fission Nuclear Reactors are fueled by uranium or plutonium, themselves finite energy sources. Spent uranium can be converted to fissile plutonium in a breeder reactor, however, a process that makes nuclear energy almost infinitely renewable. Nuclear technology, however, has not yet developed either failproof reactors or a safe method for disposing of nuclear wastes (*see* Waste Disposal). The development of nuclear fusion (whose end products are harmless) has so far been hindered by the difficulties of containing the fuels (plentiful light elements such as hydrogen) at the extremely high temperatures necessary to initiate and sustain fusion. Renewable energy sources include the energy from water and wind (see Turbine; Water Wheel; windmill); geothermal energy, the earth's internal heat that is released naturally in Geysers and Volcanoes; tidal energy, the power released by the ebb and flow of the ocean's tides; biomass, the use of certain crops (including wood) or crop wastes either directly as fuel or as a fermentable source of fuels such as alcohol or methane (*see* Gasohol); and Solar Energy, which can be stored and used directly as heat, or transformed into electricity through the use of

Photovoltaic Cells. All these renewable energy sources are presently being tapped in some form, but none can replace fossil fuels without huge advances in the technologies needed to exploit them.

Engineering. Profession devoted to designing, constructing, and operating the structures, machines, and other devices of industry and everyday life. Before the Industrial Devolution, the field was confined to the military engineer, who built fortifications and weapons, and the civil engineer, who built bridges, harbors, aqueducts, buildings, and other structures. During the early 19th cent., mechanical engineering developed to design and build manufacturing machines and the engines to power them. Major modern engineering fields include the allied pursuits of mining, metallurgical, and petroleum engineering; electrical engineering, encompassing the generation and transmission of electric power and the design of all the devices that use it; electronics engineering, dealing mainly with computers and communications equipment, chemical engineering, dealing with the design, construction, and operation of plants and machinery for making such products as acids, dyes, drugs, plastics, and synthetic rubber; and aerospace engineering, comprising the design and production of aircraft, spacecraft, and missiles. Industrial, or management, engineering is concerned with efficient production and the design of methods and processes to achieve it.

Engraving. Intaglio printing process in which the lines to be printed are cut into a metal (usually copper) plate with a graver, or burin. The earlies known engravings printed on paper date from the mid-15th cent. Early master engravers include Durer, Schongauer, and Lucas Van Leyden. During the 19th cent., steel engraving, which can produce many prints, was used to produce reproductions; it was superseded by photomechanical processes.

Enrichment. The process of increasing the concentration of one particular isotope in a mixture of isotopes.

Enthalpy. Symbol *H*. A thermodynamic property of a system equal to the sum of its internal energy (*U*) and the product of its pressure (*p*) and its volume (*V*): i.e. $H = U + pV$. Changes in enthalpy are useful in chemical thermodynamics because many chemical reactions are carried out at constant (atmospheric) pressure. The measured heat change when a chemical reaction occurs is the change in enthalpy of the system (ΔH): i.e. the sum of its internal-energy change and the work done by a change in volume. By convention. ΔH is negative for an exothermic process.

Entomology. Study of insects, an arthropod class comprising about 675,000 known species, and representing about nine tenths of all classified animal species. Insects are studied for purely biological reasons, as well as for their role as crop pollinators; carriers of viral, bacterial, fungal, and protozoal diseases; parasites of man and livestock; destroyers of economically important plants; and predators of other destructive insects.

Entropy. Quantity specifying the amount of disorder or randomness in a system bearing energy or information. In Thermodynamics, entropy indicates the degree to which a given quantity of thermal energy is available for doing useful work — the greater the entropy, the less available the energy. According to the second law of thermodynamics, during any process the change in entropy of a system and its surroundings is either zero or positive; thus the entropy of the universe as a whole tends towards a maximum. In Information Theory entropy represents the "noise," or random errors, occurring in the transmission of signals or messages.

Entropy. Symbol : *S*. A measure of the disorder of a system. Originally, the quantity was introduced in connection with the second law of thermodynamics. In any system undergoing reversible change the change of entropy is defined as the energy absorbed divided by the thermodynamic temperature: $dS = dq/T$. The entropy of the system is thus a measure of the availability of its energy for performing useful work. When a practical (irreversible) change occurs, the entropy of a close system always increases.

In fact the entropy of a system, defined in this way, is a measure of the way in which the total energy of the system is distributed amongst its constituent atoms. The statistical idea of entropy has been extended to include changes that do not involve changes in energy. In general the entropy of a system is a measure of its degree of "order" — the more disordered the system, the higher its entropy. For example, two pure liquids in a container will gradually mix together by diffusion. The mixture has a higher entropy than the separated substances because it is more disordered. Similarly, when a liquid freezes part of the entropy change is due to a decrease in entropy caused by a change to a more ordered arrangement of atoms in the crystal.

Enzyme. Protein functioning as a biological Catalyst. Enzymes accelerate (often by several orders of magnitude) chemical reactions in the cell that would proceed imperceptibly or not at all in their absence. The enzyme itself is not permanently modified by its participation. Most enzymes demonstrate great specificity, reacting with only one or a small group of closely related chemical compounds: thus, sometimes several enzymes are required for efficient catalytic function. Some enzymes depend on the presence of Coenzymes for their function. For the enzyme to continue to be effective, its three-dimensional molecular structure must be maintained. X-ray crystallography is used to analyze the structure of enzymes. Over 1,000 different enzymes have been identified, and the exact sequence of Amino Acids (subunits of a protein) has been determined for many proteins since 1967, when the first such determination was made. It is believed that enzymes function by attaching the substrate molecule to a specific molecular site, so that the electrostatic forces of nearby atoms sharply reduce the energy needed to cleave and re-form the appropriate chemical bonds.

Eosin (bromeosin). A red crystalline solid, $C_{20}H_3Br_4O_8$, showing a greenish fluorescence when in aqueous solution. It is used in colouring textiles and inks.

Ephedrine. Mild, slow-acting drug used to treat moderate attacks of bronchial asthma and to relieve nasal congestion from hay fever or infection of the upper respiratory tract. Nonaddictive, ephedrine may cause insomnia and restlessness.

Ephemeris. A table of calculated astronomical data, giving movements and positions of planets and other celestial bodies.

Ephemeris. Tables listing the positions of one or more celestial bodies for each day of the year. Early national ephemerides include the French *Connaissance des temps* (begun 1679) and the British *Nautical Almanac and Astronomical Ephemeris* (begun 1767). Since 1958 the U.S. and Great Britain have jointly published solar, lunar, planetary, and satellite ephemerides in identical publications (the U.S. american Ephemeris and the *British Astronomical Ephemeris).*

Epicycloid. A plane curve that is the locus of a point on the circumference of a circle as that circle rolls on the outside of another circle. *Compare* hypocycloid.

Epidemiology. Field of medicine concerned with the study of epidemics, outbreaks of disease that affect large numbers of people. Epidemiologists, using sophisticated statistical analyses, field investigations, and complex laboratory techniques, investigate the cause of a disease, its distribution (geographic, ecological, and ethnic), method of spread, and measures for control and prevention. Epidemiological investigations once concentrated on such communicable diseases as Tuberculosis, Influenza, and Cholera, but now also encompass Cancer, Heart Disease, and other diseases affecting large numbers of people.

Epidiascope. An apparatus that can function both as a slide projector and an episcope.

Epilepsy. Disorder of the brain characterized by sudden, brief, localized seizures or by widespread convulsive seizures involving loss of consciousness and motor control. The development of epilepsy may be idiopathic (of unknown cause) or symptomatic (with the cause identified). Infantile epilepsy may result from developmental abnormalities, metabolic disease, or birth injury. In adults, causes of epilepsy include trauma, tumors, and organic brain disorders. Drug therapy includes the use of anticonvulsants such as phenytoin and phenobarbital.

Epimerism. A form of optical isomerism in which a molecule has two or more asymmetric atoms and two isomers *(epimers)* differ in the arrangement about one of these atoms. The term is used in carbohydrate chemistry to describe optical isomerism resulting from different arrangements of the groups attached to the carbon atom next to the aldehyde group.

Epinephrine or **adrenaline.** Hormone secreted by the medulla of the Adrenal Glands. Strong emotions, such as fear and anger, produce an increase in heart rate, muscle strength, blood pressure, and glycogen metabolism, causing epinephrine to be released into the bloodstream and prolonging the acceleration of such bodily functions. See also Catecholamine.

Epiphyte or **Air plant.** Plant that grows not in the soil but on another plant, depending on it only for physical support; an epiphyte makes its own food, thus differing from a Parasite. Epiphytes obtain moisture from the air or moisture-laden pockets of the host plant. Well-known examples are many Orchids and Bromeliads.

Episcope. An optical apparatus for projecting an image of a flat opaque picture, diagram, etc., into a screen. The object is illuminated with a high-intensity light and a real image is projected by large lenses. *See also* epidiascope.

Epitaxy. The growth of a layer of one solid substance on the surface of another in such a way that the deposited material has the same crystal structure as the underlying substrate. *Epitaxial layers* can be formed by condensation from the vapour: the method is extensively used in making junctions in integrated circuits and other semiconductor devices.

Epoxyethane. *See* ethylene oxide.

Epoxy Group. The group -O- linked to atoms that are themselves linked together, thus forming a ring.

Epoxy Resin. Any of a class of synthetic resins made by polymerization of compounds containing epoxy groups. Epoxy resins are viscous liquid substances containing either linkages (-O-) and epoxy groups. They can be mixed with certain curing agents, such as amines or anhydrides, to produce a hard clear resistant plastic. Epoxy resins are also used as adhesives, the resin and the curing agent being mixed together before being applied.

Epsom salts. A water-soluble, bitter-tasting compound ($MgSO_4.7H_2O$) that occurs as white or colorless needleshaped crystals. It is used as a medicinal purgative, in leather tanning, in mordant dyeing, and as a filler in cotton goods and paper. It is found in waters of mineral springs and as the mineral epsomite.

Equation. 1. A mathematical statement that two expressions have the same value. The equation may hold for particular values of the variables, as in the equation $x^2 + 3x + 2 = 0$ which holds only for $x = -2$ and $x = -1$. Alternatively it may hold for all values of x, as in $x^2 + 3x + 2 = (x + 1)(x + 2)$, in which case it is an identity. Equations in science are mathematical statements of relationships between physical quantities.

2. A statement, usually in symbols, that two quantities or mathematical expressions are equal, e.g., $x + 3 = 6 - y$ or $\sin^2\theta = 1 - \cos^2\theta$. A numerical equation contains only numbers, e.g., $2 + 3 = 6 - 1$; a literal equation contains some letters, representing unknowns or variables. An identity is an equation that is true no matter what value is substituted for the variables, such as $x^2 - 1 = (x - 1)(x + 1)$; a conditional equation is true for only certain substitutions, e.g., only the values +1 and -1 make true the equation $x^2 - 1 = 0$. See also Algebra; Root.

3. A symbolic representation of a chemical reaction using chemical formulae. The right-hand side of the equation represents the products; the left-hand side the reactants. For example, the equation $2NaOH + H_2SO_4 = Na_2SO_4 + 2H_2O$ indicates that two molecules (or moles) of sodium hydroxide combine with one of sulphuric acid to yield one molecule of sodium sulphate and two of water. The numbers of each type of atom are the same on both sides of the equals sign: the equation is then said to be *balanced*. In chemistry a reaction that proceeds completely is often represented using a single arrow (—>). One that represents a chemical equilibrium uses the double arrow (=).

Equation of state. An equation relating the pressure, volume, and temperature of a pure substance, these three quantities serving to define the state of the substance. *See* gas laws, van der Waals' equation.

Equator. Imaginary great circle around the earth, equidistant from the two geographical poles and forming the base line from which Latitude is reckoned. Measuring c.24,000 mi (38,600 km), the equator intersects N South America, central Africa, and Indonesia. See also Astronomical Coordinate Systems.

Equilateral. Designating a plane figure with all its sides equal in length.

Equilibrium. 1. (in mechanics). The state of a body or system when it has no resultant force or resultant couple acting on it. Its condition is then unchanging: usually the term is applied to a body or structure that is stationary under the influence of balanced forces. Three types of mechanical equilibrium are distinguished depending on the behaviour when the system is subjected to a small momentary displacement from its equilibrium position. In *stable equilibrium,* the system regains its original position. An example of a stable system would be a ball inside a spherical bowl: the ball always rolls back to the lowest point. *Unstable equilibrium* occurs if the initial displacement causes the system to move spontaneously further from the initial position. A ball balanced on top of a sphere, for example, would roll off the sphere if displaced. In *neutral equilibrium* the system takes up a new position as a result of the displacement: an example is that of a ball placed on a plane surface.

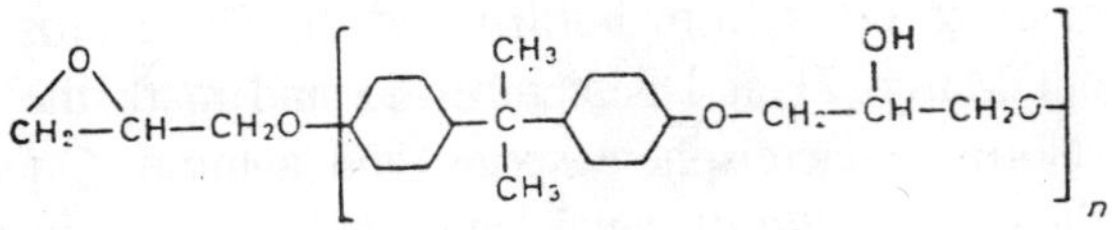

Fig. E-8. Epoxy polymer.

2. (in physics and chemistry). In general, any state of a system in which the properties are not changing with time. A reversible chemical reaction, for example, may reach a condition in which the concentrations of reactants and products are unchanging: the rate of the forward reaction equals that of the back reaction (*see* chemical equilibrium). This is an example of *dynamic equilibrium*—a state in which opposing processes occur at the same rate. Another example is that of a saturated vapour, in which molecules escape from the liquid into the vapour phase at the same rate as they condense. *Thermodynamic equilibrium* is the condition of a system when its free energy is a minimum. Both the above examples involve states of thermodynamic equilibrium. However, not all unchanging systems are in this state. A supersaturated vapour or solution, for instance, does not have a minimum free energy and can change spontaneously be precipitating liquid or solid. Such systems are said to be in *metastable equialibrium.*

Equinox. One of two points at which the ecliptic crosses the celestial equator. In the Northern Hemisphere, the *vernal equinox* occurs on about March 21 and the *autumnal equinox* occurs on about September 23. At these times the days and nights have equal length. The terms are used both for the points of intersection and for the times at which they occur. Compare solstice.

Equinox. Either of two points on the celestial sphere where the Ecliptic and the celestial equator intersect (see Astronomical Coordinate Systems). The vernal equinox is the point at which the sun appears to cross the celestial equator from south to north; the autumnal equinox, the point where it appears to cross from north to south. These crossings occur about March 21 and September 23 and mark the beginnings of Northern Hemisphere spring and autumn. On either date night and day are of equal length (12 hr each) in all parts of the world. The equinoxes are not fixed points, but move westward along the ecliptic (see Precession of the Equinoxes).

Equipartition of energy. The principle that the total energy of molecules is equally distributed amongst their degrees of freedom. In a gas, the average energy per degree of freedom is $kt/2$, where k is the Boltzman constant and T the termodynamic temperature. Thus a monatomic gas, which has tree translational degrees of freedom, has an energy per molecule of $3kT/2$.

Equipotential. Denoting a line or surface connecting points with the same electric potential.

Equivalent circuit. An electrical circuit containing simple resistors, inductors, and capacitors, that is identical in properties to a more complicated electrical device.

Equivalent proportions, law of. *See* reciprocal proportions.

Equivalent weight. The number of grams of hydrogen that an element could combine with or displace in a chemical reaction.

Equivalent weights are sometimes called *chemical equivalents.* They are numbers assigned to substances to indicate the proportions (by weight) in which they combine or react together, using hydrogen as a standard. If an element does not form compounds with hydrogen its equivalent weight can be found by taking the mass that combines with or displaces 8 grams of oxygen or 35.5 grams of chlorine. In general, the equivalent weight of an element is equal to its atomic weight divided by its valency. The equivalent weight of an acid is its molecular weight divided by the number of acidic hydrogen atoms per molecule.

Erbium. Symbol: Er. A soft silvery metallic element belonging to the lanthanide series. A.N. 68; A.W. 167.26; m.pt. 1522°C; b.pt. 2510°C; r.d. 9.05; valency 3.

Erbium. (Er), Metallic element, discovered by Carl G. Mosander in 1843 in the mineral gadolinite. A silvery, malleable Rare-Earth Metal, erbium is in the Lanthanide Series. The rose-colored oxide erbia is used as a coloring agent in glazes and glass.

Erecting prism or **lens.** A prism or concave lens used to produce an erect image from an inverted image in a terrestrial telescope or other optical instrument.

E-region. *See* ionosphere.

Erg. A CGS unit of energy equal to the work performed when a force of one dyne is moved through a distance of one centimetre in its direction of action. It is equivalent to 10^{-7} joule.

Ergot. Disease of rye and other cereals caused by a fungus *Claviceps purpurea),* which appears on heads of rye as dark purple structures called ergots. The ergots, which resemble rye seeds, contain certain active substances (Alkaloids) that are poisonous—ergot poisoning was epidemic in the Middle Ages, for example, because bread was often made from ergot-contaminated rye. The ergots also contain alkaloids used as drugs: e.g., ergotamine to treat migraine headaches and ergonovine to stop postpartum uterine bleeding.

Erlenmeyer flask. A conical laboratory flask with a narrow neck and flat broad bottom. [After Emil Erlenmeyer (1825-1909), German chemist].

Erosion. General term for the processes by which the surface of the earth is constantly worn away, principally by the abrasive action of running water, waves, Glaciers, and wind Streams and ocean waves, for example, erode bedrock by their own impact or by the abrasive action of the debris they carry. Rock surfaces are eroded by glaciers moving over them or by wind driving sand and other particles against them. In the U.S. the washing away of farmland topsoil is a problem. Erosion is also caused by strip mining and the removal of vast areas of plant cover, e.g., by careless land developers. Some methods of preventing soil erosion are reforestation, terracing, and special plowing techniques.

Esaki diode. *See* tunnel diode.

Escape velocity. The minimum velocity that would have to be given to an object for it to escape from a specified gravitational field. The escape velocity from the earth is 11 200 metres per second.

Escape velocity. The velocity that a body must be given in order to escape the gravitational hold of some other larger body, e.g., the earth, moon, or sun Escape velocity depends on the mass of the larger body and the distance of the smaller body from its center. The escape velocity from the earth's surface is about 7 mi/sec (11.3 km/sec).

Essential oil. 1. A volatile oily liquid mixture extracted from a plant. Essential oils are often fragrant substances containing terpenes and esters.

2. Volatile oils that occur in plants and in general give the plants their characteristic odors, flavors, or other such properties. Generally complex mixtures of organic compounds, they are used in perfumes, flavorings, and medicines. Among the plants notable for their essential oils are members of the following plant families: Carrot (e.g., Anise, Dill, angelica). Ginger (cardamom), Laurel (Cinnamon, Camphor), Mint (peppermint, Thyme), Myrtle (Clove), and Orchid (Vanilla). See also Fats and Oils.

Ester. Any of a class of organic compounds produced by reaction of acids with alcohols. For example, acetic acid and ethyl alcohol give ethyl acetate; $CH_3COOH + C_2H_5OH = CH_3COO\text{-}C_2H_5 + H_2O$.

This reaction is called *esterification:* it is reversible and good yields of ester and obtained by distilling off the product as it is formed or by removing the water in some way. The esters of simple acids and alcohols are often volatile fragrant compounds. Glycerides are naturally occurring esters of glycerol and fatty acids.

Esters can be hydrolysed to give the original acid and alcohol—the reaction is the reverse of esterification and is sometimes called saponification.

Estrogen. Any of a group of hormones synthesized by the reproductive organs and adrenal glands. Estrogens are important in the regulation of the female menstrual cycle (see Menstruation).

Estuary. Partly enclosed coastal body of water, open to the ocean so that fresh and salt water are mixed. Estuaries are extremely sensitive and ecologically important habitats, providing waterfowl sanctuaries and breeding and feeding grounds for many desirable life forms. They are often excellent harbors, and most large U.S. ports are located in them. The human impact on estuaries (e.g., landfill, sewage pollution, industrial effluents) can destroy the very properties of the estuary that facilitated development of the region.

Etalon. An interferometer in which interference is caused by multiple reflections of light between two parallel semi-reflecting surfaces.

Eta meson. Symbol : Gη. A type of meson having zero charge and a mass 280.5 times that of the electron.

Etching. Intaglio process and the print produced from it. A metal plate, usually copper or zinc, is coated with a ground of acid-resistant resin or wax. The design is drawn on the ground with a needle, exposing the metal. Then the plate is submerged in acid, cutting into the exposed lines. The plate is removed frequently, and lines that are sufficiently deep are coated with varnish; others remain longer and produce darker, heavier lines when printed. In printing, varnish is removed and the plate is warmed, coated with ink, wiped so that ink remains only in the grooves, covered with moist paper, and run through a press. Artists often alter etching plates, and several states of a work may exist. The art of etching evolved from Engraving; both apparently originated in Germany. Among the artists foremost as etchers are Durer, Callot, Rembrandt, Tiepolo, Piranesi, Goya, and Whistler.

Ethanal. *See* acetaldehyde.

Ethane. A colourless odourless flammable gaseous alkane. C_2H_6, which occurs in natural gas. It is used in organic synthesis and as a fuel. M.pt. -183°C; b.pt. -89°C; r.d. 1.04 (air = 1).

Ethanedioic acid. *See* oxalic acid.

Ethanediol (ethylene glycol). A colourless hygroscopic viscous liquid, CH_2OHCH_2OH, with a sweet taste. It is made by hydration of ethylene oxide and used as an antifreeze, humectant, and solvent. M.pt. -13°C; b.pt. 197°C; r.d. 1.1.

Ethanenitrile. *See* acetonitrile.

Ethanoic anhydride. *See* acetic anhydride.

Ethanol (alcohol, ethyl alcohol). A colourless volatile flammable liquid, C_2H_5OH. Ethanol is the common alcohol found in intoxicating drinks, in which it is prepared by fermentation. On an industrial scale, ethanol is manufactured by catalytic hydration of ethene or by fermentation of molasses. It has many uses, particularly as a solvent and as a raw material for the manufacture of other chemicals. M.pt. -117°C; b. pt. 78°C; r.d. 0.81.

Ethanol or Ethyl alcohol. (CH_3CH_2OH), a colourless liquid with characteristic odor and taste, commonly called grain alcohol or, simply, Alcohol. Ordinary ethanol is about 95% pure, the remaining 5% being water, which can only be removed with difficulty to give pure or absolute ethanol. Ethanol is the alcohol in beer, wine, and liquor, and can be made by the Fermentation of sugar or starch. Denatured alcohol, for industrial use, is ethanol with toxic additives. Ethanol is used as a solvent in the manufacture of varnishes and perfumes; as a preservative; in medicines; as a disinfectant; and as a fuel. Ethanol is a soporofic; if its presence in the blood exceeds about 5%, death usually occurs. Behavioral changes, impairment of vision, or unconsciousness occur at lower concentrations. See Alcoholism.

Ethanoyl chloride. *See* acetyl chloride.

Ethene (ethylene). A colourless flammable gaseous alkene, C_2H_4, with a sweet taste and odour. It is manufactured by cracking petroleum gases and is used in the preparation of many polymers. M.pt. - 169°C; b.pt. - 102°C; r.d. 0.61 (liquid at 0°C).

Ethenoid resin. Any synthetic resin produced by polymerizing compounds that contain double bounds. Common examples are polythene, polypropylene, the vinyl resins, and the acrylic resins.

Ether. 1. *See* diethyl ether.

2. Any of a class of chemical compounds with the general formula ROR', where R and R' are alkyl or aryl groups.

3. (aether). An elastic weightless fluid formerly thought to permeate all space and to be the medium in which light and other electromagnetic waves are propagated by vibration. *See also* Michelson-Morley experiment.

Ethnology. Scientific study of the origin and functioning of humans and their cultures, usually considered a branch of cultural Anthropology. In the 19th cent. historical ethnology attempted to explain extant cultures by examining their early development. In the 20th cent. the comparative study of cultures has predominated.

Ethology. Study of animal behaviour, especially its physiological, ecological, and evolutionary aspects. Originally, an organism's actions were classified as either instinctive behavior (actions not influenced by the animal's previous experiences, e.g., common reflexes) or learned behavior (actions dependent on earlier experiences, e.g., problem solving). Current emphasis is on the interaction between environmental and genetically determined responses, particularly during early development.

Ethyl acetate (ethyl ethanoate). A Colourless fragrant flammable liquid ester, $CH_3COOC_2H_5$. It is used in flavourings and perfumery and as a solvent for lacquers and plastics. M.pt. -83°C; b.pt. 77°C; r.d. 0.89.

Ethyl alcohol. *See* ethanol.

Ethylamine (aminoethane). A colourless volatile flammable liquid amine, $C_2H_5NH_2$, made by reacting chloroethane with alcoholic ammonia. It is used in making dyes and as a stabilizer for rubber latex. M.pt. -81°C; b.pt. 16.6°C; r.d. 0.69 (15°C).

Ethyl benzene. A colourless flammable liquid derivative of benzene, $C_6H_5C_2H_5$, made by a Friedel-Craft's reaction. It is used in the manufacture of styrene. M.pt. -95°C; b.pt. 136°C; r.d. 0.8.

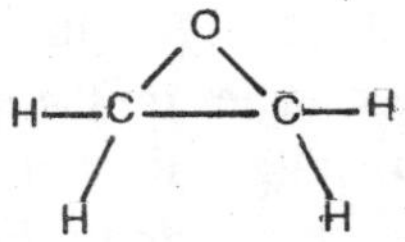

Fig. E-9. Ethylene oxide.

Ethylene oxide (epoxyethane). A colourless flammable gaseous ether, C_2H_4O, used in the manufacture of detergents and acrylonitrile. M.pt. -111°C; b.pt. 11°C; r.d. 0.87.

Ethyl ethanoate. *See* ethylacetate.

Ethyl group. The organic group C_2H_5-.

Eucalyptus. Evergreen shrubs or trees (genus *Eucalyptus)* of the Myrtle family, a characteristic component of Australian flora. Also called ironbark, bloodwood, and gum tree, it is the sole food source of the Koala and is valued for its timber and Essential Oils. Several eucalyptus species are among the tallest trees known; *E. amygdalina regnans* reaches a height of over 300 ft (91 m).

Euclid. Fl. 300 B.C., Greek mathematician whose treatment of elementary plane Geometry serves as the basis for most beginning courses on the subject. His great contribution was the use of a deductive system of proof in his *Elements,* a presentation of the mathematics of his day in 13 books. Studies of his work in the 19th cent. gave rise to several types of Non-Euclidean Geometry.

Euclidean geometry. Set of propositions in geometry that can be derived by rigorous logical steps from the five postulates stated by Euclid at the beginning of his *Elements.* The first three postulates state that it is possible (1) to draw a straight line from any point to any point, (2) to produce a finite straight line continuously in a straight line, and (3) to describe a circle with any center and radius. The fourth postulate states that "All right angles are equal to one another," and the fifth posits that "If a straight line falling on two straight lines makes the interior angles on the same side less than two right angles, the two straight lines, if produced indefinitely, will meet on that side on which are the angles less than two right angles." A Non-Euclidean Geometry replaces the fifth postulate with either of two alternative postulates.

Euclidean geometry. Geometry based on the assumptions made by Euclid. In his book *Elements* a large number of theorems concerning the properties of figures are deduced from a number of definitions, of concepts such as point and line, and axioms (or postulates). Together the definitions and axioms are considered to be self-evident. Euclid's fifth postulate, the "parallel postulate," says that if a point lies outside a line it is possible to draw only one line through the point that is parallel to the first line. Euclidean geometry is the geometry usually applied to physical measurements involving distance and angle. However, forms of non-Euclidean geometry exist, in which the parallel postulate is not necessary, and these may be more applicable under special circumstances. [After Euclid (c. 300 B.C.) Greek mathematician.]

Eudiometer. An apparatus for measuring volume changes at constant pressure during chemical reactions either in the gas phase or in the solid and liquid phases when evolution of gaseous products occurs. Information about the stoichiometry and in some cases about the mechanism of the reactions can be obtained from these measurements.

Eudoxus of Gnidus. Yoodok'ses), 408?-355? B.C., Greek astronomer, mathematician, and physician. He was the first Greek astronomer to explain the movements of the planets scientifically, holding that a number of concentric spheres supported the planets in their paths. It is claimed that he calculated the length of the solar year, indicating a calendar reform like that made later by Julius Caesar, and that he was the discoverer of some parts of geometry included in Euclid's work.

Euglenophyta. Small division of the plant kingdom consisting of photosynthetic, aquatic organisms. Most are unicellular; many have flagella and are motile. They resemble plants in that they have Chloroplasts, but they also have gullets and lack cell walls, like animals. The most characteristic genus is *Euglena,* common in ponds and pools.

Euler, Leonhard. (oi'ler), 1707-83, Swiss mathematician. The most prolific mathematician who ever lived, he worked at the St. Petersburg Academy of Sciences, Russia (1727-41, 1766-83), and at the Berlin Academy (1741-66). Euler contributed to numerous areas of both pure and applied mathematics, including calculus, analysis, number theory, topology, algebra, geometry, trigonometry, analytical mechanics, hydrodynamics, and the theory of the moon's motion.

Euler's theorem. *See* polyhedron. [After Leonhard Euler (1707-83). Swiss mathematician].

Europium. Symbol; Eu. A soft ductile silvery-while metallic element belonging to the lanthanide series. A.N. 63; A.W. 151.96; m.pt. 822°C; b.pt. 1597°C; r.d. 5.25; valency 2 or 3.

Europium. (Eu), metallic element, purified by Eugene Democracy in 1901. A silvery-white Rare-Earth Metal, it has physical properties like those of other Lanthanide Series members, but its chemical properties resemble those of calcium. Europium is used in nuclear-reactor control rods.

Eutectic. Mixture of two substance in such proportions that the mixture solidifies as a whole when the liquid is cooled, forming a solid solution. In general, if two liquids are mixed and the mixture cooled one or other of the components will solidify until the liquid remaining has the eutectic composition, at which point the whole mixture solidifies. This is the lowest liquid temperature attainable— the *eutectic point*. Cryohydrates and many alloys are eutectics.

Euthanasia. Either painlessly putting to Death (positive euthanasia) or failing to postpone death from natural causes (negative euthanasia) in cases of terminal illness. The term *negative euthanasia* has come to mean withdrawal of extraordinary means (e.g., intravenous feeding, respirators, and artificial kidney machines) used to preserve life. Positive euthanasia is illegal in the U.S., but physicians may lawfully refuse to prolong life when there is extreme suffering. Debate centers around whether measures used to keep the terminally ill alive are routine, and justifiable, or extraordinary. Those seeking to legalize positive euthanasia assert that a patient has a right to decide to refuse treatment even though that may result in death. The problem of incapacitation can be solved, it is held, by a living will, in which a person expresses, in writing, preferences concerning medical treatment and death.

Eutrophication. Aging of a lake or slow-moving stream by biological enrichment of its water. In a young lake, the water is cold and clear, supporting little life. With time, plant and animal life burgeon, and organic remains begin to be deposited. As the lake grows shallower and warmer, marsh plants take root and begin to fill in the basin. Eventually the lake gives way to bog, finally becoming dry land. The natural aging of a lake may span thousands of years. However, wastes from human activities can accelerate the aging process, as with water Pollution. The prime pollutants are nitrates and phosphates, which greatly stimulate the growth of algae, producing a pungent surface scum. The decomposition of dead algae reduces the water's dissolved oxygen content, adversely affecting fish and other aquatic life forms typical of a mature lake.

Evans, Oliver. 1775-1819, American inventor; b. near New port, Del. Evans developed and patented a number of grain-handling machines, including an elevator and a conveyer system, that later became standard equipment in U.S. mills. He was a pioneer in the design of high-pressure steam engines, and in 1804 he built the first land vehicle in the U.S. that moved under its own power, an engine-equipped dredge.

Evolution. Concept embodying the belief that existing organisms descended from a common ancestor. This theory, also known as descent with modification, constitutes organic evolution. Inorganic evolution deals with the development of the physical universe from unorganized matter (see Cosmology). Organic evolution conceives of life as having begun as a simple, primordial protoplasmic mass from which arose, through time, all subsequent living forms. The first clearly stated theory of evolution, that proposed by Jean Lamarck in 1801, included the inheritance of Acquired Characteristics as the operative force in evolution. Subsequently (1858), Alfred Russel Wallace and Charles Darwin independently set forth a scientifically credible theory of evolution based on Natural Selection, focusing on the survival and reproduction of those species best adapted to the environment. This theory had a profound effect on scientific thought and experimentation. Although it has undergone modification in light of later scientific developments, the theory of evolution still rests on essentially the same grounds emphasized by Darwin, supported by research in Genetics as well as by comparative anatomy, embryology, geography, paleontology, and, recently, biochemistry. It has been challenged by those believing in the creation theory of the universe (see Creationism). See also Mutation.

Evaporation. Conversion of a liquid into a vapour, especially when the liquid is below its boiling point. The term is also applied to the process of boiling the solvent from a solution in order to obtain the solid residue.

Even-even nucleus. A nucleus that has an even number of protons and of neutrons.

Even-odd nucleus. A nucleus that has an even number of protons and an odd number of neutrons.

Evolute. A curve that is the locus of the centres of curvature of points on another specified curve (the involute).

Ewing, (William) Maurice. 1906-74, American geophysicist; b. Lockney, Tex. Ewing's oceanographic work was critical to the acceptance of the theory of Plate Tectonics. In the late 1930s he conducted the first seismic studies (1935) and the first photographing (1939) of the ocean floor. At Columbia Univ. he helped to found (1949) the Lamount-Doherty Geological Observatory and was (1949-72) its first director.

Excitaton. 1. Change of the energy of an atom, ion, molecule, etc., from one energy level (usually the ground state) to a higher energy level. The system is then in an *excited state.*

2. Production of a magnetic field by passing a current through the coil of an electromagnet, transformer, or similar device.

Excition. An electron in an excited state in a semiconductor considered as an electron bound electrostatically to the positive hole. The exciton may migrate through the lattice until the hole and electron recombine.

Exclusion principle. *See* Pauli exclusion principle.

Excretion. Process of eliminating from an organism the waste products of metabolism and other materials of no use. In one-celled organisms wastes are discharged through the cell's surface. Higher plants eliminate gases through the stomata, or pores, on the leaf surface, and multicellular animals have special excretory organs. In humans the main excretory organs are the Kidneys and other organs of the Urinary System, through which Urine is eliminated, and the large intestine, from which solid wastes are expelled (see Digestive System). Excretion also takes place in the Skin, which eliminates water and salt in the form of sweat, and the Lungs, which expel water vapour and carbon dioxide.

Exobiology. Search for extraterrestrial life within the solar system and throughout the universe. Philosophical speculation that there might be other worlds similar to ours dates back to the ancient Chinese and Greeks. Six basic parameters determine whether an environment is suitable for life as we know it: temperature, pressure, salinity, acidity, water availability, and oxygen content. Advanced life is restricted to a narrow range of these parameters, but primitive micro organisms can exist over a much wider range. Data already collected by Space Probes essentially rule out advanced life on other planets of our solar system. Some scientists, however, have estimated that as many as 50,000 planets in our galaxy have earthlike conditions and that a substantial fraction of these are likely to have cultures as technologically advanced as our own; efforts to detect radio emissions from these civilizations have so far been unsuccessful. A continuing effort is being made to detect primitive life within our solar system, e.g., the surface landings (1976) of the Viking space probes.

Exoergic. *See* exothermic.

Exosphere. The highest region of the earth's atmosphere, extending from a height of about 400 kilometres.

Exothermic. Denoting a process or chemical reaction that takes place with the evolution of heat. Nuclear processes that evolve heat are said to be *exoergic*. *Compare* endothermic.

Expanding universe. The universe thought of as expanding in size, as shown by the red shift of distant celestial objects.

Expansion. An increase in the size of a solid, liquid, or gas as a result of an increase in temperature. The expansion results from increased motion of the atoms or molecules. The opposite process—*contraction*— occurs when the solid is cooled. The tendency of a substance to expand is given by its *coefficient of expansion,* α, which is the fractional increase in size per unit rise in temperature; i.e. $\alpha = \Delta s/st$, where *s* is the original size. Δs the increase, and *t* the temperature rise. A solid has three coefficients of expansion reflecting increases in length (linear expansion), area (superficial expansion), and volume (volume expansion). Linear expansion obeys the equation: $l_t = l_o (1 + at)$. *where* l_0 is the original length, l_t the length at a certain temperature, and *t* the *rise* in temperature. Similar equations give the area coefficient, β, and the volume coefficiert, γ. Coefficients of expansion of solids are measured by the comparator method or by techniques involving the movement of interference fringes.

Liquids have two coefficients of volume expansion: a coefficient of *apparent expansion,* in which the expansion of the container is not corrected for, and a coefficient of *absolute* or *real expansion,* which is based on the true expansion of the liquid. The apparent coefficient is the sum of the absolute coefficient and the volume coefficient of the material of the container. Apparent coefficients can be measured by a dilatometer and absolute coefficients by Dulong and Petit's method. The expansion of gases is measured at constant pressure and for ideal gases it is 1/273 of the original volume per degree.

Explosive. A substance that undergoes rapid decomposition or combustion, evolving much heat and producing a large volume of gas. The heat evolved causes the gas to expand greatly, exerting the enormous pressure of explosions. Important explosives include TNT (trinitrotoluene), Dynamite, and Nitroglycerin. Chemical explosive are of two general kinds. Some, e.g., Gunpowder, are mixtures of readily combustible but not necessarily explosive substances, which, when set off (by ignition), undergo very rapid combustion. Others, called high explosives (e.g., TNT), are compounds whose molecules are unstable and can undergo explosive decomposition (detonation) without burning. The latter kind is used in warfare (e.g., in bombs, explosive shells, torpedoes, and missile warheads) and for blasting rock in mining and construction work. Nondetonating explosives, e.g., gunpowder and modern smokeless powders, are used as propellants for bullets and in fireworks. Nuclear explosive release energy by transformation of the atomic nucleus (see Atomic Bomb; Hydrogen Bomb; Nuclear Energy).

Exponent. The number indicating the power to which something is raised. Thus in a^n, n is the exponent. Exponents satisfy the laws: $a^n a^m = a^{n+m}$, $(a^n)^m = a^{nm}$, $(ab)^n = a^n b^n$, and $1/a^n = a^{-n}$.

Exponent. In mathematics, a number or algebraic expression written above and to the right of another number or algebraic expression called the base. In the expressions x^2, and x^n, the number 2 and the algebraic letter n are the exponents, respectively, of the base x. A positive whole number as exponent indicates the power, or how many times the base is to be taken as a factor (e.g., $2^3 = 2 \times 2 \times 2 = 8$). Fractional and negative exponents indicate, respectively, Roots and reciprocals (e.g., $8^{1/3} = 3\sqrt{8}$ and $x^2 = 1/x^2$). A zero exponent makes any quantity equal to 1.

Exponential function. A mathematical function e^x , or Ae^{ax}, where A and a are constants. The irrational number e is the base of natural logarithms and has a value of 2.718... The function e^x is sometimes written exp x. It can be expanded as the infinite series $l + x + x^2/2!$.... An *exponential law* is one expressed in the form Ae^{ax}. Thus, expoential decay of a quantity y with time follows an equation of the form $y = Ae^{-axt}$.

Extender. Any of various substances added to paints, synthetic adhesives, and rubbers to dilute them (for economy) or to modify their physical properties.

Extraordinary ray. *See* double refraction.

Extremely high frequency (EHF). A frequency in the range 30 gigahertz to 300 gigahertz.

Extrinsic Semiconductor. *See* semiconductor.

Eye. 1. A section of the human eye is shown in the illustration. Light enters through the transparent *cornea* and is focused by the *crystalline lens* onto the *retina,* the light-sensitive layer of nerve cells on the interior of the eye. The retina is composed of two types of nerve cell—*rods,* which only detect low intensities of light and do not sense different colours, and *cones,* which are the cells detecting normal light intestities. The cones are responsible for colour vision: there are thought to be three types, sensitive to red, green, and blue light and colour sensations are produced by a combination of responses. The most efficient part of the retina is a small region with a high density of cones, the *fovea centralis.* The *blind spot* is a small area with no light sensitivity at the point where the *optic nerve* enters the eye. Besides the retina, the eye has two other layers: the *choroid* and the outer protective *sclera.*

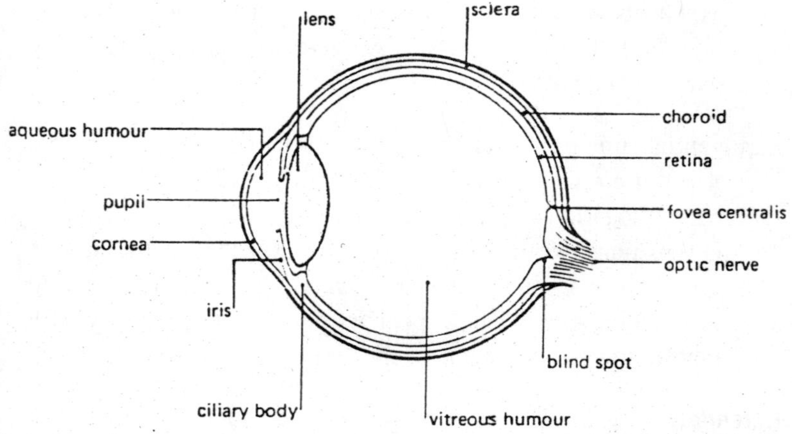

Fig. E-10. Eye

The *iris* is an arrangement of muscle tissue in front of the lens forming a hole, the *pupil,* through which light passes. This increases or decreases in size according to the light intensity. Two fluids are present in the eye: the *aqueous humour* is a water solution between the cornea and the lens and the *vitreous humour* is a gelatinous material behind the lens.

Under normal conditions the lens is held under tension by *cilliary muscles* and focused on distant objects. To observe nearer objects the muscles relax, causing the lens to bulge and shorten its focal length. This process is known as *accommodation.* Defects of vision include hypermetropia, myopia, presbyopia, and astigmatism.

2. Organ of vision. Like a camera, it has a diaphragm and variable focusing. The eyeball has three covering layers. The *sclera,* the outermost layer, is partially visible as the "white" of the eye. In the center of the sclera and projecting slightly is the cornea, a transparent membrane that acts as the window of the eye. The *choroid,* the layer lying beneath the clear, is composed of dense pigment and blood vessels. Near the center of the visible part of the eye the choroid forms the ciliary body, the muscles of which change the shape of the lens. The ciliary body merges with the iris, a muscular diaphragm that regulates the size of the pupil, the round opening through which light enters the eye. The iris, where it is not covered by the sclera, reveals the choroid's pigmentation, usually brown or blue (giving the eye its color). Behind the iris is the lens, a transparent, elastic, but solid ellipsoid body that bends light rays, focusing them on the *retina,* the third tissue layer. The retina is a network of nerve cells, notably the rods and cones, that send impulses along the optic nerve to the brain. The rods provide vision in dim light, while the cones respond best to bright light and provide color vision. See also Astigmatism, Blindness; Color Blindness; Farsightedness, Glaucoma; Nearsightedness.

Eye lens. *See* eyepiece.

Eyepiece (eye lens). The lens or lens system that is nearest the eye in a telescope, microscope, or other optical instrument. It is used to view the image formed by the objective.

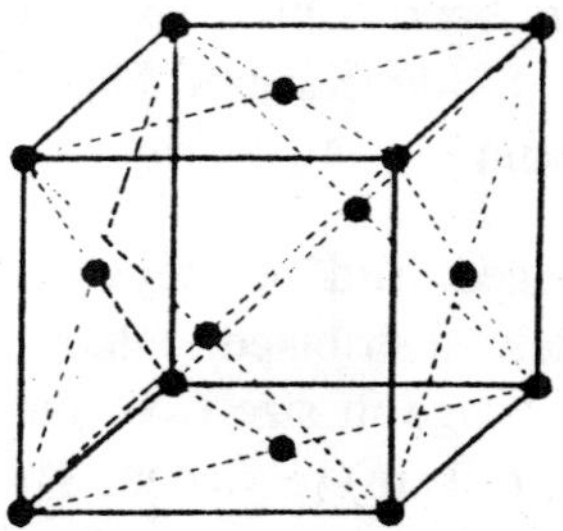

Fig. F-1. Face-centred cubic unit cell.

Face-centred. Denoting a crystal structure in which the unit cell has eight atoms at its corners and a further six atoms arranged at the centres of each face. In a *face-centred cubic* lattice (*see* illustration) the unit cell is a cube. *Compare* body-centred.

Factor. Any of two or more integers that give a specified number when multiplied together. Thus, 3 and 5 and 15 and 1 are factors of 15.

Factorial. A mathematical function of a numbers equal to the product of the integers from 1 to the number. Thus, factorial n is $n(n-1)n-2)$...3.2.1. It is written $n!$ or n. y definition, O! = 1.

Faculae. Bright irregular patches on the sun's disc that are much brighter and hotter than the general surroundings. They are apparently associated with sunspots.

Fahrenheit scale. A temperature scale in which the melting point of ice is 32° and the boiling point of water is 212° at standard pressure. Temperature is measured in *degrees Fahrenheit* (°F). [After Gabriel Daniel Fahrenheit (1686-1736), German physicist.]

1. A covalent bond is more likely to be produced if the ions have large charges.

2. A covalent bond is more likely if the cation is small or the anion large. [After Kasimir Fajans (b. 1887). U.S. chemist born in Poland.]

Falcon. Long-winged Bird of prey (genus *Falco)* related to the Hawk. Widely distributed, falcons are typified by notched beaks. They range in size from the $6^1/_2$-in. 16.5-cm) falconet to the 24-in. (60-cm) gyrfalcon. The commonest and smallest American falcon is the sparrow hawk *(F. sparverius).* Falcons lay their eggs on the ground, on cliff ledges, or in deserted hawk or Crow nests. The peregrine falcon and gyrfalcon are used in falconry.

Fall-out. Radioactive material settling on the earth's surface from the atmosphere as a result of a nuclear explosion.

Family. A set of curves, surfaces, etc., whose equations are all obtained from one another by replacement of a certain constant by other constants. For example $y = 3x + a$ represents a family of straight lines given by different values of the constant a.

Family therapy. From of Psychotherapy dealing with the improvement of family relationships and the emotional environment. Unlike Group Psychotherapy, which concentrates on the individual within a group, family therapy focuses on the emotional life of the family as a whole. By defining the existing family structure and clarifying patterns, conflicts, and alliances, it enables family members to develop alternative patens of relating. Treatment may include both family and individual sessions with one or a team of therapists.

Farad. Symbol : F. The SI unit of electric capacitance, equal to the capacitance of a parallel-plate capacitor with a potential difference of one volt across its plates and holding a charge of one coulomb. The unit is too large for practical use;; the microfarad is more commonly employed. [After Michael Faraday (179-1867), English scientist.]

Faraday. *See* Faraday's laws of electrolysis

Faraday constant. *See* Faraday's laws of electrolysis.

Faraday disc. A simple direct current generator consisting of a metal disc rotated in a steady magnetic field. The e.m.f. is induced between the centre of the disc and the edge, the current being drawn by brushes at these points.

Faraday effect. Rotation of the plane of polarization of a beam of plane-polarized light in certain media as a result of the application of a strong magnetic field parallel to the direction of the light. The amount of rotation depends on the nature of the medium, the path length, and the field strength. The effect was originally discovered with light but is also observed for other types of electromagnetic radiation.

Faraday, Michael, 1791-1867. English scientist. Despite little formal education he laid the foundations of classical Field theory, later fully developed by James Clerk Maxwell. Faraday worked (1813-62) at the laboratory of the Royal Institution in London, becoming its director in 1825. He developed the first dynamo (in the form of a copper disk rotated between the poles of a permanent magnet), the precursor of modern dynamos and Generators. From Faraday's and Joseph Henry's independent discoveries (1831) of electromagnetic Induction stemmed a vast development of electrical machinery for industry. In 1825 Faraday discovered the compound Benzene. He formulated (1834) Faraday's law, which states that the number of moles of substance produced at an electrode during Electrolysis is directly proportional to the number of moles of electrons transferred at that elecrode.

Faraday's laws of electrolysis. Two laws used to express the amount of chemical change produced during electrolysis:

1. The amount of chemical change produced is proportional to the quantity of electricity passed.

2. The amount of chemical change produced in substances by a fixed quantity of electricity is proportional to the equivalent weight of the substance.

It follows from Faraday's laws that the charge Q required to deposit or dissolve n moles of an ion with valency v is given by $Q = Fnv$. F is a constant (the *Faraday constant*) having the value 9.648 670 X 10^4 coulombs per mole.

The amount of electricity required to deposit or dissolve one mole of a monovalent ion is 9.648 670 X 10^4 coulombs. This is sometimes called the *faraday* and used as a unit of charge.

Far Sight. *See* presbyopia.

Farsightedness. or hyperopia, defect of vision in which near objects appear blurred, but far objects can be seen clearly. Because the eyeball is too short or the eye's refractive power too weak, the image is focused behind the retina of the Eye rather than upon it. Eyeglasses with convex lenses can compensate for the refractive error (see Lens).

Fast neutron. A neutron with a high kinetic energy, usually taken to be one with an energy greater than 0.1 MeV. The term is also used for neutrons with sufficient energy to cause fission in a fast nuclear reactor: i.e. those with energies above 1.5 MeV.

Fast Reactor. *See* Fission reactor.

Fat. Any of various mixtures of glycerides found in plant and animal tissues, distinguished from natural oils by the fact that they are solid at room temperature.

Fathom. A unit of length equal to six feet. It is usually used in expressing the depth of the sea.

Fault. In geology, a fracture in the earth's crust in which the rocks on each side move in relation to each other. This movement can be vertical, horizontal, or oblique. Horizontal faults showing lateral displacement of points originally directly opposite each other, such as the San Andreas Fault in California, are called strike-slip. Another type of horizontal fault, the transform fault, occurs on the ocean floor between offset portions of the mid-ocean ridge. Altered positions of rocks due to faulting can mislead the geologist, because one rock layer can appear twice in the same cross section or disappear entirely.

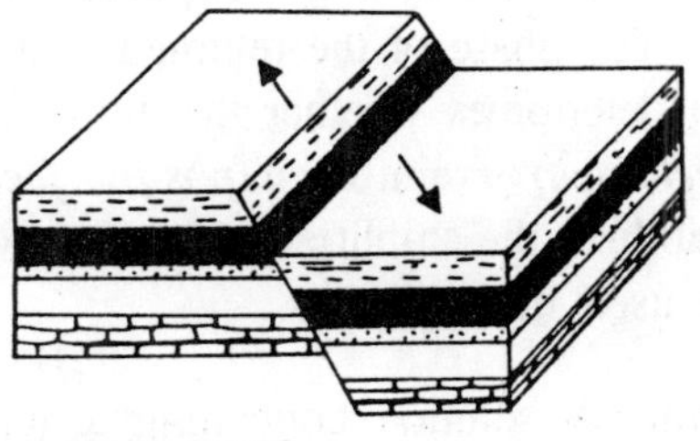

NORMAL FAULT

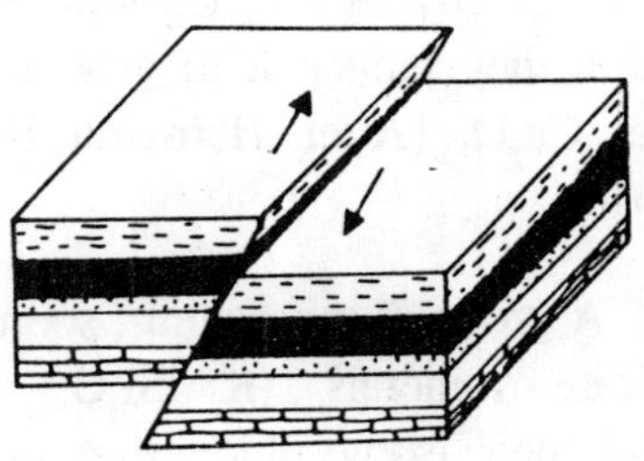

REVERSE FAULT

Fig. F-2. Type of vertical faults.

Feathers. Outgrowths of the skin that constitute the protective and decorative plumage of birds and are thought to have evolved from reptilian scales in Mesozoic times. Feathers grow only along defined tracts; full grown, they lack a blood supply and have hollow shafts. Typically, barbs radiate from the distal part of the shaft (the rachis) and interlock via smaller cross-linking barbules to form a web, which gives the feather flexibility and durability. Down feathers of young birds and protective undercoats of aquatic birds lack webs. Specializations (e.g., crests, ruffs, and topknots) and modifications (e.g., bristles) exist. Most feather colors (e.g., red, yellow, brown) are due to pigment in the feather, some (e.g., green and violet), as well as iridescent effects, are due to reflection and diffraction of light. The most important roles of feathers are in flight and heat retention.

Feedback. Any process in which the output of a device or system controls the input in some way. The term is commonly applied to the coupling of a portion of the output of electronic amplifiers to the input. The phase of the returned signal relative to that of the input determines whether the amplification is reduced (*negative feedback*) or reinforced (*positive feedback*). Negative feedback stabilizes the amplifier and cuts down noise. Positive feedback is used in oscillators.

Fehling's Solution. A solution containing a mixture of copper sulphate, sodium hydroxide, and potassium sodium tartrate, used as a test for aldehydes and ketones. When heated with Fehling's solution they reduce it to give a red precipitate of copper I oxide, Cu_2O, [After Hermann Fehling (1812-85), German chemist.]

Feldspar or **Felspar.** A group of potassium-, sodium-, and calcium-aluminum silicate minerals ($KAlSi_3O_8$, $NaAlSi_3O_8$, and $CaAl_2Si_2O_8$) and their isomorphic mixtures. The three pure members are called, respectively, orthoclase, albite, and anorthite. As constituents of crystalline rocks, the feld-spars form much of the earth's crust. Pure feldspar is colorless and transparent, but impurities commonly make it opaque and colorful. Potassium feldspars are used in making porcelain and as a source of aluminum in making glass. The plagioclase feldspars (those ranging in composition from albite to anorthite) are commonly gray and occasionally red. Moonstone is a gem variety.

Femto. Symbol: f. Prefix indicating (10^{-15}).

Fennel. Common name for several herbs, particularly those of the genus *Foèniculum* of the Carrot family. Their licoricescented foliage and seeds are used for flavoring. Sweet fennel has a thickened leaf base and is eaten like Celery.

Fermat, Pierre de. (ferma), 1601-05 French mathematician and magistrate. Although his work in mathematics was done for recreation, he was a founder of modern Number Theory and probability theory. Many developments in number theory have resulted from unsuccessful attempts to prove Fermat's Last Theorem, a still unproved conjecture that states that the equation $x^n + y^n = z^n$, where x,y, and z are nonzero integers, has no solutions for n when n is an integer greater than 2.

Fermat's Principle of least time. The principle that a ray of light (or other electromagnetic radiation) moves between any two points in such a way that the time taken is a minimum: i.e. less than the time that would be taken for other possible paths. [After Pierre Fermat (1601-65), French mathematician.]

Fermentation. A reaction in which compounds such as sugar are broken down by the action of microorganisms that form enzymes which catalyse the reaction. A common fermentation reaction is the conversion of sugar to alcohol: $C_6H_{12}O_6 = 2C_2H_5OH + 2CO_2$. The reaction is catalysed by the enzyme *zymase.* Industrial fermentation processes using various yeasts, bacteria, and moulds are used to produce such compounds as butanol, glycerol, acetic acid, and citric acid.

Fermentation. Anaerobic degradation of glucose and other simple sugars in the living cell. Of the many different kinds of glucose fermentation, two kinds predominate; one, in microorganisms and animal cells, produces lactic acid as the sole end product; the other, in brewer's yeast and some bacteria, yields ethyl alcohol and carbon dioxide. Alcoholic fermentation to produce intoxicating beverages was practiced in antiquity. By 1500 B.C. production of beer from germinating cereals (malt) and of the wine from crushed grapes was an established technical art in most of the Middle East. The modern science of biochemistry emerged directly from 19th-cent. studies of fermentation.

Fermented Milk. Whole of skim milk curdled and thickened by lactic-acid-producing microorganisms. It was used in many forms by early nomadic herdsmen, especially in Asia, S and E Europe, Scandinavia, Africa, and South America. Yogurt, acidophilus milk, cultured buttermilk, and kumiss (koumiss) are forms of fermented milk.

Fermi. A unit of length equal to 10^{-15} metre. It is used in nuclear and atomic physics. [After Enrico Fermi (1901-54), Italian-American physicist.]

Fermi-Dirac statistics. A type of statistical mechanics applied to the distribution of particles amongst energy levels when the particles are indistinguishable from each other and only one particle can occupy each energy state.

Fermi, Enrico. 1901-54, American physicist; b. Italy; came to U.S., 1938. He contributed to the early theory of beta decay and the Neutrino and to quantum statistics and discovered the element Neptunium. For his experiments with Radioactivity he was awarded the 1938 Nobel Prize in physics. Fermi created (1942) the first self-sustaining chain reaction in uranium and worked on the atomic bomb at Los Alamos. He later helped develop the hydrogen bomb and served on the General Advisory Committee of the U.S. Atomic Energy Commission.

Fermi Level. The highest occupied energy level in a solid.

Fermi National Accelerator Laboratory. (Fermilab), physical science research center (est. 1968) located near Batavia, Ill. Universities Research Association operates it under contract to the U.S. Dept. of Energy. Work at Fermilab is devoted to the study of elementary particles, principally through the use of a synchrotron particle accelerator capable of accelerating protons up to energies of 500 billion electron volts.

Fermion. Any particle that obeys Fermi-Dirac statistics. All baryons and leptons, such as the proton and the electron, are fermions. *Compare* boson.

Fermium. Symbol: Fm. A transuranic actinide element made by neutron bombardment of plutonium. The most stable isotope^{252}, Fm, has a half-life of about 23 hours. A.N. 100.

Fermium (Fm). Radioactive element, discovered in 1952 by Albert Ghiorso and colleagues in residue from a thermonuclear explosion. Its physical properties are largely unknown; its chemical properties are similar to those of other Actinide Series members.

Fern. Any plant of the division Polypodiophyta, consisting of several thousand species found worldwide, usually in tropical rain forests. Most common living ferns belong to the polypody family (polypodiaceae) and are characterized by triangular fronds subdivided into many leaflets (pinnae) and smaller pinnules. Except for ornamentals, the only commercially important ferns are tree ferns (families Dicksoniaceae and Cyatheaceae), whose trunks are used in construction; the starchy pith is used as stock food. Ancestors of modern ferns were the dominant vegetation during the Carboniferous era. They, and such relatives as Club Mosses and Horse-tails, are the most primitive plants to have developed a true vascular system. Ferns reproduce by alternation of generations, and although no present-day ferns reproduce by seed, there is some fossil evidence that fernlike plants are the ancestors of seed plants (conifers and true flowering plants).

Ferrate. Any salt containing the ion FeO_4^{2-}. Ferrates are salts of the hypothetical acid H_2FeO_4 (*Ferric acid*).

Ferric. *See* iron.

Ferric Acid. *See* ferrate.

Ferricyanide. Any salt containing the complex ion $[Fe(CN)_6]^3$, in which the iron is in the + 3 oxidation state. Ferri-cyanides are salts of the hypothetical acid $H_2Fe(CN)_6$ (*ferricyanic acid*).

Ferrimagnetism. A type of magnetic behaviour in which the material has a low magnetic susceptibility that increases with temperature. It is found in certain inorganic solids, such as ferrites, and resembles antiferromagnetism with the difference that neighbouring antiparallel magnetic moments are unequal.

Ferrite. 1. Any of a class of compounds with the formula MFe_2O_4, where M is a divalent transition metal. The ferrites are mixed oxides. $MO.Fe_2O_3$, rather than salts and they are generally ceramic materials exhibiting ferromagnetism or ferrimagnetism. Although they are magnetic they are nonconductors of electricity, a combination that makes them useful in transformer cores and other equipment liable to suffer power loss from eddy currents. Ferrites are also used in computer storage units.

2. Iron with its body-centred cubic crystal structure, existing either as the pure element or as a constituent of steels.

Ferroalloy. Any alloy of iron and one or more other metals made by smelting mixtures of the metal ore with iron ore. Ferroalloys are used to introduce other metals into steel.

Ferrocene. An orange crystalline solid, $(C_5H_5)_2Fe$, made from iron II chloride and cyclopentadienyl sodium (Na^+ - C_5H_5). Ferrocene was the first sandwich compound to be prepared. M.pt. 173°C.

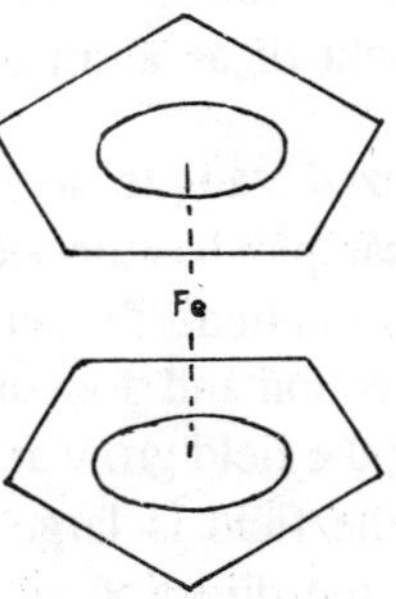

Fig. F-3. Ferrocene.

Ferrochromium. An alloy of iron and chromium (50-70%) used in making steels.

Ferrocyanide. Any salt containing the complex ion $[Fe(CN)_5]^4$, in which the iron is in the +2 oxidation state. Ferrocyanides are salts of the hypothetical acid $H_4Fe(CN)_6$ *(ferrocyanic acid).*

Ferroelectricity. A property of some dielectrics of exhibiting a permanent electric polarization that is maintained in the absence of a polarizing field. Ferroelectric materials include Rochelle salt and barium titanate. The effect is caused by the presence of electric dipoles in the crystal which line up in domains, just as magnetic dipoles are aligned in the domains of a ferromagnetic material. A graph of electric displacement against applied electric field has a hysteresis loop similar to ferromagnetic hysteresis.

Ferromagnetism. A type of magnetic behaviour in which the material has a very high susceptibility that depends on temperature. Some ferromagnetic substances, such as steel, are also capable of remaining in a magnetized state in the absence of an external magnetizing field (i.e. they form permanent magnets). Ferromagnetism is caused by unpaired electrons, as in paramagnetism. These act as small elementary magnets and in ferromagnetic materials they are aligned parallel to each other within regions of the solid called *domains* by intermolecular forces known as *exchange forces*. Each domain can thus be thought of as a small magnet.

In an unmagnetized sample the domains are orientated at random so the sample has no net magnetic moment. If an external field is applied, the elementary magnets tend to align along its direction and domains whose magnetic moments are directed along the field grow at the expense of neighbouring domains. When the field is large enough, all the elementary magnets point in the direction of the field and the sample is *saturated.* Permanent magnetism occurs when the elementary magnets remain aligned in the absence of the field. Ferromagnetism is found in iron, cobalt and nickel, and in certain of their compounds.

Ferromanganese. An alloy of iron and manganese (70-80%) used in making steels.

Ferrosilicon. An alloy of iron and silicon (up to 15%) used in making steels.

Ferrosoferric oxide. *See* Iron oxide.

Ferrotungsten. All alloy of iron and tungsten (60-85%) used in making steels.

Ferrous. 1. *See* Iron.

2. Containing or indicating metallic iron.

Ferrovanadium. An alloy of iron and vanadium used in making steels.

Fertile. Denoting a material that can be transformed into fissile material for use in a fission reactor.

Fertility drug. Any of a variety of substances used to increase the potential for conception and successful pregnancy by correcting various functional disorders. In the male, inadequate sperm production can often be remedied by administering testosterone and sometimes pituitary hormones. In the female, failure to ovulate, the most common cause of female sterility, can sometimes be treated with the administration of gonadotropic hormone and, in some cases, clomiphene citrate. Often inducing more than one ovum in a month, the use of clomiphene may result in multiple births. In cases of repeated miscarriage or bleeding during pregnancy, Progesterone has been found effective.

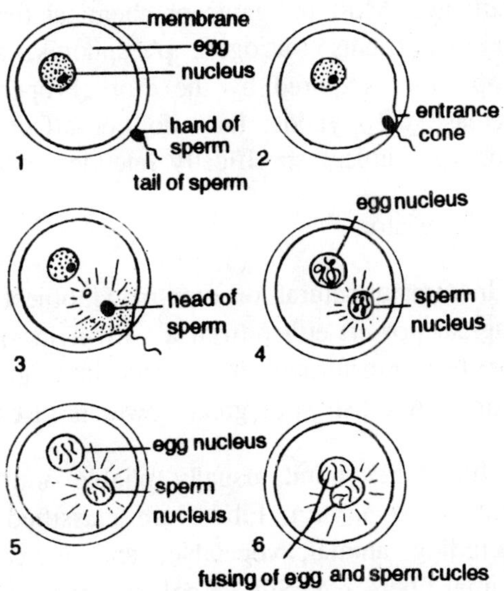

Fig. F-4. Fertilization of an egg cell.

Fertilization. In biology, sexual reproductive process involving the union of two unlike sex cells (gametes)—the ovum, or egg (female), and sperm (male)—followed by fusion of their nuclei. The principle of fertilization is the same in all organisms. The ovum absorbs the first sperm to make successful contact and the two nuclei fuse, combining the hereditary material of both parents; the subsequent Embryo develops into a new individual. In lower plants and animals, the sperm swims to the egg through an external medium or through fluid in the female reproductive tract. In some higher plants, Pollination enables the sperm to contact the egg. In higher animals, sperm contact initiates cell division in the fertilized egg (zygote).

Fertilizer. Organic or inorganic material added to the soil to replace or increase plant nutrients. Organic fertilizers—including animal and green manure, fish and bone meal, guano (seabird exrement), and Compost—are decomposed by soil microorganisms, and their elements are freed for plant use. Most inorganic or chemical fertilizers contain the major nutrients (nitrogen, phosphorus, and potassium) in proportions required by the crop. Properly used, fertilizers increase crop yields; they do not affect a crop's nutritive properties unless specifically intended to do so.

Fet. *See* transistor.

Fetish. Inanimate natural or man-made object believed to have magical power, either from a will of its own or from a god that has transformed the object into an instrument of its desires. A fetish with great power is often declared Taboo.

Fiber. Threadlike strand, usually pliable and capable of being spun into a yarn. Fibers are classified as either natural (including animal, vegetable, and inorganic) or artificial. Animal fibers, e.g., Silk, Wool, and goat hair (mohair), consist mainly of proteins. Vegetable fibers, e.g., Cotton, consist mainly of cellulose. Asbestos is the chief natural inorganic fiber, Fiber Glass is also of inorganic origin. Artificial fibers are made by either synthesizing Polymers, as in nylon, or altering natural fibers, as in Rayon. Fibers are used for textiles, e.g., cotton, silk, wool; cordage, e.g., hemp; brushes, e.g., animal hairs; filling, e.g., horsehair; and plaiting, e.g., sisal. The invention of Spinning and Weaving machinery during the Industrial Revolution greatly increased the demand for fibers.

Fiber glass. Thread made from glass. Molten glass is forced through a kind of sieve, thus spinning it into threads. Strong, durable, and impervious to many caustic and to extreme temperatures, fiber glass fabrics are widely used in industry. Some, resembling silk and cotton, are used for drapery. Boat hulls and auto bodies molded of fiber glass combined with plastic are rustproof.

Fiber optics. Transmission of messages or information by light pulses along hair-thin glass fibers. Cables of optical fibers can be made smaller and lighter than conventional cables using copper wires or coaxial tubes, yet they can carry much more information, making them useful for transmitting large amounts of data between computers and for carrying data-intensive television pictures or many simultaneous telephone conversations. Optical fibers are immune to electromagnetic interference (from lightning, nearby electric motors, and similar sources) and to crosstalk from adjoining wires, and tapping into them is more easily detected. To keep a signal from deteriorating, optical fibers requires fewer repeater over a given distance than does copper wire. In addition to communications, optical fibers are starting to be used in automobiles and aircraft and are expected to have many other applications.

Fibonacci numbers. The sequence of numbers each of which is the sum of the two preceding. It begins: 1, 1, 2, 3, 5, 8, 13, 21, ...(After Leonardo Fibonacci of Pisa, c. 1170-1250).

Fibre optics. Optical techniques for transmitting images along flexible transparent fibres. Under certain conditions, light falling on the end of a glass fibre is guided along it by a series of total-internal reflections. Bundles of such fibres can be used for viewing or photographing inaccessible objects, inspecting machine parts, medical diagnosis, etc.

Field. A region in space in which a force is exerted on an object as a result of its charge (electric field), magnetic dipole moment (magnetic field); or mass (gravitational field). Thus a charged particle can be thought of as producing an electric field in space so that another particle experiences a force at points in the field. Similarly a magnet or a current in a wire creates a magnetic field that can influence a magnetic dipole. A field is often visualized as *lines of force* through space. The direction of a line at a point is the direction of the force and the density of lines indicates the strength of the field.

Field. In physics, region throughout which a force may be exerted; example are the gravitational, electric, and magnetic fields that surround, respectively, masses, electric charges, and magnet (see Gravitation; Electricity; Magnetism). Field are used to describe all cases in which two bodies separated in space exert a force on each other.

Field-effect transistor. *See* transistor.

Field emission. Emission of electrons from the surface of a solid as a result of the existence of a high electric field at the surface. The effect is also called *cold-cathode emission.* It is used in the *field-emission microscope,* a device in which a high negative voltage is applied to a fine metal point placed at the centre of a spherical fluorescent screen. Electrons emitted from the metal are accelerated to the screen in straight lines, producing a magnifield image of the metal tip. The device has to be used at very low pressures of gas.

Field ionization. Ionization of atoms or molecules at the surface of a solid under the influence of a high electric field. Electron transfer occurs between the molecule and the solid to produce a positive ion. The *field-ion microscope* is a similar device to the field-emission microscope with the difference that the metal point has a high positive voltage and a low pressure of helium gas is used. The image is produced by helium ions hitting the screen. The device has a very high resolution: individual atoms can be observed.

Field magnet. *See* electric motor.

Fig. Plant genus *Ficus)* of the Mulberry family, comprising over 600 species of vines, shrubs, and trees. Fig plants bear hundreds of tiny female flowers inside a fleshy receptacle, which ripens into a soft, pear-shaped fruit containing masses of tiny seeds. The common fig (*F. carica)*, native to the Mediterranean region and cultivated from early times for its commercially valuable fruit, has been naturalized elsewhere in mild, semiarid climates. Some edible varieties can be pollinated only by the fig wasp.

Filler. Any insert substance added to a resin, rubber, or similar substance either to dilute it (for economy) or to modify its physical properties.

Film badge. A piece of photographic film in a plastic holder worn by workers exposed to radiation. The amount of blackening of the film indicates the radiation dose received: different types of radiation are detected by the use of metal filters.

Filter. 1. A device for separating suspended solid particles from a liquid or gas by passing the fluid through a porous medium such as paper, sand, sintered glass, etc.

2. A device for passing light or other electromagnetic radiation with a restricted range of wavelength. Light is filtered by use of pieces of coloured glass or film.

3. An electrical circuit that allows passage of alternating currents within a certain range or frequency while stopping currents that have frequencies outside the allowed range.

Filter pump. A type of vacuum pump in which a jet of water forced through a narrow nozzle traps air molecules and removes them from the system (see illustration). It cannot produce pressures below the vapour pressure of water.

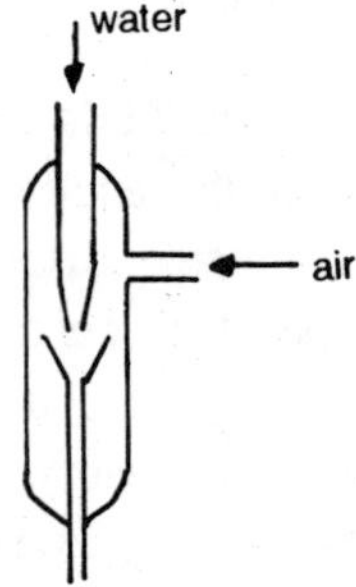

Fig. F-5. Filter Pump.

Filtrate. The clear liquid obtained by filtration.

Filtration. The process of separating solids from a fluid using a filter.

Finch. Bird of the family Fringillidae, found worldwide except Australia, considered the most highly developed of birds. Finches are typified by stout, conical bills, used to open the seeds that comprise their main diet; many also eat insects. Highly diversified, finches are classified in three groups: those with triangular bills, e.g., Canary, Sparrow, and the birds called finch, e.g., bullfinch and goldfinch; those with thick, rounded bills, e.g., Cardinal; and those whose mandibles cross at the tips, e.g., crossbill. Because their seed diet does not depend on weather, many finches are year-round residents in colder areas.

Finder (collimator). A small telescope attached to the side of a large astronomical telescope to enable the large telescope to be directed towards the object to be observed.

Fine structure. Structure observed in a spectral line or band when it is viewed at high resolution. Thus a line in an emission spectrum may prove to be a number of closely spaced lines caused by the spin of the electron. *Hyperfine structure* is similar but even higher resolution is required to direct it; the term is used to disignate closely spaced lines produced by emission or absorption in atoms of different istopes or splitting caused by interaction of the spin of the nucleus with that of the electron.

Fingerprint. An impression of the underside of the end of a finger or thumb, which has ridges that from a pattern unique to each person. Fingerprinting as an identification device dates from antiquity, but sophisticated modern methods are based on work by the English scientist Sir Francis Galton and other during the late 19th cent. Fingerprint identification is used extensively in criminal investigation, the armed service, government employment, and banking. In the U.S. a national fingerprint file is maintained by the Federal Bureau of Investigation.

Fir. Tree (genus *Abies*) of the Pine family, found chiefly in the alpine regions of the Northern Hemisphere. Tall, pyramidal evergreens, firs have short, flat, stemless needles and erect, cylindrical cones. They are valued for their fragrance and beauty. The balsam fir is a popular Christmas tree.

Fire damp. Methane formed in coal mines. It forms an explosive mixture with air. *See also* afterdamp.

Firefly or **lightning bug.** Small, luminescent, nocturnal, carnivorous Beetle of the family Lampyridae. They emit heatless flashes of green-yellow to red-orange light (see Bioluminescence); these are sexual signals produced by special organs on the abdomen. Firefly larvae prey on snails and earthworms, and some are beneficial to agriculture. Larvae and wingless females are called glowworms.

First aid. Immediate and temporary treatment of a victim of sudden illness or injury while awaiting the arrival of medical aid. Essential or first aid treatment include correct bandaging of wounds, application of splints for fractures and dislocations, artificial respiration, and treatment of Shock, bites and stings, and Burns. Some conditions, e.g., severe bleeding, cessation of breathing, and poisoning, require immediate attention to prevent death.

Fischer-Tropsch process. A process for producing various hydrocarbons, alcohols, or aldehydes by catalysed reactions between carbon monoxide and hydrogen. [After Franz Fischer (1852-1932), German chemist, and Hans Tropsch (1889-1935), Czech chemist.]

Fischer-Tropsch Process. Method, discovered (1923) by the German coal researchers Franz Fischer and Hans Tropsch, for the synthesis of hydrocarbons and other aliphatic compounds. A mixture of hydrogen and carbon monoxide is reacted in the presence of an iron or cobalt catalyst. Much heat is evolved, and such products as methane, synthetic gasoline and waxes, and alcohols are made, with water or carbon dioxide by-products.

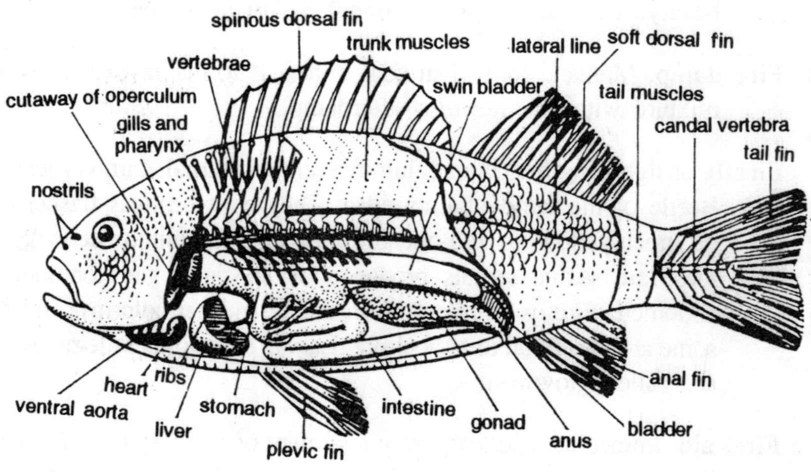

Fig. F-6. Anatomy of a ray-finned fish.

Fish. Limbless, aquatic, Vertebrate animal with fins, internal gills, and skin with a glandular secretion that decreases friction. Most fish have scales and are cold-blooded. A typical fish is torpedo-shaped, with a head containing a brain and sensory organs, a muscular-walled trunk with a cavity containing internal organs, and a muscular tail. Most fish propel themselves through the water with weaving movements and control their direction with fins. Although some fish, such as sharks, bear live young, most fish eggs are fertilized and hatch in water. There are over 20,000 species in three classes: Agnatha, the most primitive fishes, jawless and without paired pelvic and pectoral fins (e.g., Lamprey) Chondrichthyes, cartilaginous fishes, with skeletons of cartilage but no swim bladder or lungs (e.g., Shark, Ray); and Osteichthyes, bony fishes, the most highly developed. Bony fishes have a bony skeleton and a swim bladder or lungs, and are divided into fleshy-finned (e.g., Lobefin, Lungfish) and ray-finned fishes (e.g., Bass, Tuna). The largest fish is the whale shark *(Rhincodon typus)*, reaching 50 ft (15m) in length and found worldwide in tropical seas; the smallest is the 1/2-in. (1.3 - cm) goby (family Gobiidae) of the Philippines.

Fissile. Capable of undergoing nuclear fission. The term is sometimes restricted to materials that undergo fission by the impact of slow neutrons.

Fission. The process in which an atomic nucleus splits into fragment nuclei of comparable size. Nuclear fission is usually accompanied by emission of neutrons and gamma rays and the production of large amounts of energy. It may be spontaneous or induced by the impact of particles or photons. *See also* nuclear weapon, fusion.

Fission bomb. *See* nuclear weapon.

Fission reactor. A device for producing energy by nuclear fission. In general a fission reactor contains a *core* of fissile material in which a chain reaction occurs by neutron capture. The reaction is regulated by *control rods* — rods of cadmium, boron, or some other neutron-absorber that can be moved in or out of the core to control the number of neutrons. The heat generated is taken away by a *coolant,* such as liquid sodium, flowing through channels in the core. It is used to heat water to drive a steam turbine for generating electricity. In the *thermal reactor* the fuel is natural uranium, which is mainly the heavy nonfissile uranium-238 with a small amount of fissile uranium-235. Uranium-238 nuclei capture neutrons to produce stable uranium-239 and under normal circumstances a chain reaction would not be sustained. A moderator is used in the core to slow the neutrons to thermal neutrons, which are less likely to be captured by the uranium 238 isotope. In this way a self-sustaining chain reaction can be maintained by the lighter nuclei. In a *heterogeneous reactor* the core is composed of thin fuel rods embedded in the moderator. A *homogeneous reactor* has a core in which the moderator and fuel are mixed. *Boiling-water reactors* are thermal reactors in which water is used as both moderator and coolant. In *gascooled reactors,* the coolant is carbon dioxide.

The other main type of fission reactor — the *fast reactor* — does not have any moderator. Instead the chain reaction occurs by capture of fast neutrons. For this to be possible the fuel must be enriched, either with uranium-235 or with plutonium-239. The core is surrounded by a *blanket* of natural uranium, in which some neutrons cause the conversion of uranium-238 to plutonium-239, which is used to enrich the fuel. Such devices are called *breeder reactors.*

Fitch, John, 1743-98. American inventor; b. Windsor, Conn. An early experimenter with steam engines and steamboats (see Steamship), he is believed to have designed (1786) the first practical steamboat. Nevertheless, he failed to receive either the opportunity to commercialize his invention or the recognition he justly deserved.

Fitting reaction. *See* Wurtz reaction. [After Rudolph Fittig (1835-1910), German chemist.]

Five-Year Plan. A program to speed indsutrialization and the collectivization of agriculture in the Soviet Union, initiated (1928-33) by joseph Stalin. Now a standard feature of the Soviet economy, such plans have been adopted elsewhere in the Soviet block and in developing nations.

Fixation. *See* nitrogen fixation.

Fixed point. A reference point of defined value on a temperature scale.

Fixed star. A true star distinguished from a planet such as Venus.

Flame. The hot luminous gas in which chemical reaction is occurring in flames are usually complex, involving free-radical chain reactions. Positive and negative ions are also produced. Light is evolved by fluorescence of excited molecules or ions or, in highly luminous flames, by incandescence of small solid particles of carbon produced in the reactions.

Flash photolysis. A technique for investigating the spectra and reactions of free radicals. The sample , usually in the form of a low-pressure gas or vapour, is contained in a glass or quartz tube and subjected to a brief powerful flash of light. Photolysis of the substance occurs, leading to the formation of atoms and free radicals which can be identified by their absorption spectra using an independent source of radiation. The kinetics of very fast reactions are investigated by observing the change in intensity of spectral lines with time using an oscilloscope.

Flamingo. Large pink or red tropical wading Bird (order Ciconiiformes), similar to the related Heron and Stork but with webbed feet and a unique, down-bent bill. Ranging in height from 3 to 5 ft. (90 to 150 cm), the flamingo has a long neck and legs and a broad wingspan.

Flash Point. The temperature to which a substance must be heated before it can be ignited. This is the temperature at which the sample evolves a flammable vapour.

Flatfish. Member of the order Pleuronectiformes, bottom-living Fish of the Atlantic and Pacific. Adult flatfish have an unusual, flattened body with both eyes on one side of the head. They lie with the blind, generally whitish, side on the ocean bottom; the eyed, colored side faces up. Soles are generally small, warm-water flatfishes with small eyes and mouths and few or no teeth. Flounders, generally larger, are known for changing color to match the background. The European sole (*Solea solea*) and many flounders, including the flukes (genus *Paralichthys*) and the large, voracious, cool-water halibuts (genus *Hippoglossus*), are important food fish.

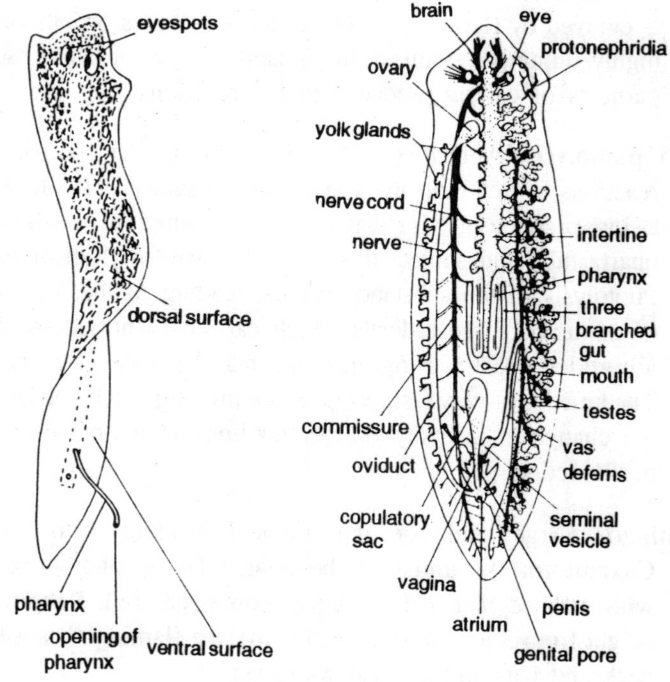

Fig. F-7. Flatworm: External (left) and internal (right) anatomy of a planaria, a representative flatworm.

Flatworm. Soft-bodied, bilaterally symmetrical invertebrate of the phylum Platyhelminthes. Among the most primitive organisms, flatworms are divided into the free-living, primarily aquatic class of turbellarians (including Planarians) and the exclusively parasitic classes of Flukes and Tapeworms.

Flax. Common name for an annual herb of the Linaceae family, especially members of the genus *Linum,* and for the fiber obtained from such plants. Native to Eurasia, flax was the major source of cloth fiber until the growth of the Cotton industry (c.1800). The flax of commerce is obtained from several varieties of *L. usitatissimum.* Exposure to water or dew and sun loosens the fiber from the woody tissue in a process known as retting. After washing, drying, beating, and combing, fibers are obtained for use in making fabric, threads, and cordage. Flax seeds are crushed to make Linseed Oil, and the remaining linseed cake is used for fodder. Dried seed is also used in medicines.

F-layer. *See* Ionosphere.

Flea. Small, blood-sucking wingless Insect of the order Siphonaptera. Adult fleas eat only blood and are external Parasites of mammals and birds. They have hard bodies flattened from side to side, piercing and sucking mouthparts, and strong legs for jumping. Certain rat fleas carry Typhus and bubonic Plague, and other fleas transmit Tapeworms to humans.

Fleming's Rules. Rules used for remembering the relative directions of motion, magnetic field, and current in dynamos and electric motors. The *left-hand rule* is used for motors. The thumb, fore-finger, and second finger are pointed at right angles to each other. If the fore-finger points in the direction of the field and the second finger points in the direction of the electric current, then the thumb indicates the direction of motion of the conductor. The *right-hand rule,* which is similar, is used for electrical generators. [After Sir John Ambrose Fleming (1849-1945), British physicist.]

Fleming, Sir Alexander. 1881-1955, Scottish bacteriologist. He discovered Penicillin (1928) and lysozyme (1922), an antibacterial substance found in body secretions. Professor of bacteriology at the Univ. of London, he shared the 1945 Nobel prize in physiology or medicine with Ernst Chain and Sir Howard Florey for work on penicillin. Fleming was knighted in 1944.

Flicker Photometer. A type of photometer in which a screen is illuminated alternately by the test source and a standard source in rapid succession. The distances of the sources are adjusted until the screen does not appear to flicker: the intensities of illumination are then equal. The instrument is particularly useful for comparing sources with different colours.

Flight. Sustained, self-powered motion through the air, as accomplished by an animal, aircraft, or rocket. Adaption for flight is highly developed in birds and insects. Birds fly by means of semicircular flapping strokes, the wings moving backward on the upstroke and forward on the downstroke. That motion pushes air downward and to the rear, creating a lift and forward thrust. Besides flapping, some birds also use gliding and soaring techniques in flight. In gliding, a bird holds its outstretched wings relatively still and relies on its momentum to keep it aloft for short distances. In soaring, a bird uses rising warm air currents to give it lift. Airships and Balloons owe their ability to ascend and remain aloft to their inflation with a gas lighter than air. Aircraft, which are heavier than air, are able to remain aloft because of forces developed by the movement of the craft through the air. The development of Jet Propulsion makes high speeds possible, and the rocket engine allows flight outside the earth's atmosphere and in interplanetary space.

Flint Glass. *See* Optical flint.

Flocculation. The process of coagulating into larger masses.

Flotation Process. Process for concentrating the metal-bearing mineral in an Ore. Crude ore is ground to a fine powder and mixed with water and reagents. When air is blown through the mixture, mineral particles cling to the bubbles, which rise to form a froth on the surface. The froth is skimmed off and the water and chemicals removed, leaving a clean concentrate. Among the minerals effectively concentrated by this method are sulfide and phosphate ores.

Flower. Specialized part of seed plants that contains reproductive organs. The basic floral parts (sepal, petal, stamen, and pistil) are modified leaves, typically arranged concentrically and attached at their bases to the tip of the stem. The outermost, green sepals (the calyx) encircle a whorl of usually showy, colored petals (the corolla), within which Pollen-bearing stamens surround a central ovary-bearing pistil. After fertilization, each ovule (the part that contains the egg) in the ovary becomes a Seed, and the ovary becomes the Fruit. The number and arrangement of floral parts varies greatly among groups of plants and are important bases for classification. In general, the higher a plant is on the evolutionary scale, the greater the flower's complexity and efficiency for reproduction.

Flowers of Sulphur. *See* Sulphur.

Fluid. Any substance that can flow: either a liquid or a gas.

Fluidics. The use of fluid flow through narrow pipes in an analgous way to the flow of electric current through circuits. Fluid circuits are much slower than electronic circuits but are unaffected by magnetic fields and ionizing radiation.

Fluidity. The reciprocal of the viscosity of a fluid.

Fluid Mechanics. Branch of Mechanics dealing with the properties and behavior of fluids, or substances that flow, i.e., liquids and gases. The larger part of the field is fluid dynamics (study of fluids in motion), which itself is divided into hydrodynamics (study of liquids in motion) and Aero Dynamics (study of gases in motion).

Fluke. Flatworm of the class Trematoda. Flukes, which are related to the Tapeworm, are internal and external parasites that cause many diseases. Adults cling to their hosts with sucking disks and bear an external cuticle resistant to digestion by the host. Some species of Flatfish are also called flukes.

Fluon. ® *See* Polytetrafluoroethylene.

Fluorescence. Luminescence in which light of a visible color is emitted from certain substances, called phosphors, when irradiated by electromagnetic radiation, especially ultraviolet light. Unlike Phosphorescence, the light is emitted only while the stimulation continues.

Fluorescein. A yellow-red powder, $C_{20}H_{12}O_5$, which exhibits a green fluorescence in alkaline solutions which viewed by reflected light. The solution is green when viewed in transmitted light. It is used as a tracer dye in examining water flow and as an indicator. Decomposes at 290°C.

Fluorescence. Luminescence that ceases as soon as the source of the excitation is removed. It is distinguished from phosphorescence—in which the emission persists. In scientific usage the term can either denote a luminescence that does not persist long enough to be recognized with the eye or can denote one that lasts less than 10^{-8} second. In general nonscientific usage the term is used for any luminescence under the action of light or electricity, regardless of the persistence.

Fluorescent Lamp. A lamp in which the light is produced by luminescence. In devices such as the sodium-vapour and mercury-vapour lamps, an electric current is passed through a low pressure of gas in a glass envelope and light is emitted from excited atoms produced by impact of electrons in the gas phase. This type of fluorescent lighting is also used in neon lights. Domestic fluorescent strip-lights consist of a glass gas-discharge tube that is coated inside with a thin layer of phosphor, which emits light as a result of the impact of electrons from the discharge.

Fluoridation. The addition of small quantities (a few parts per million) of a fluorine compound, normally sodium fluoride, NaF, to drinking water for the purpose of decreasing the incidence of tooth decay in the population.

Fluoride. Any of certain compounds containing fluorine. *See* Halide.

Fluorination. A chemical reaction in which one more fluorine atoms are introduced into a molecule.

fluorine. Symbol: F. A pale yellow corrosive pungent gaseous element occurring chiefly in fluorspar (CaF_2) and cryolite (Na_2AlF_6). It is made by electrolysis of fused hydrogen fluoride and used in producing uranium and in rocket fuels. Fluorine is diatomic. F_2, and is the most reactive of the elements, belonging to the halogen group. It attacks glass and has to be kept in special steel containers. A.N. 9; A.W. 18.9984; m.pt. -219.64; b.pt. -188°C; r.d. 1.7 (air = 1); valency 1.

Fluorine (F). Gaseous element, first prepared in 1886 by Henri Moissan. Fluorine is a yellowish, poisonous, highly corrosive Halogen gas. It is the most chemically active nonmetallic element and the most electronegative of all the elements. Fluorite is its chief commercial source. The addition of one part per million of soluble fluorides to public water supplies reduces tooth decay. Fluorine compounds are used in ceramic and glass manufacturing, in refrigeration and air-conditioning systems, and as aerosol propellants. Teflon is an inert, heat-resistant fluorocarbon.

Fluorite or **Fluorspar.** Calcium fluoride mineral (CaF_2), occuring in various colors in crystal, granular, and massive forms. Its crystals sometimes exhibit fluorescence. Found mainly in England, Germany, Mexico, Kentucky, and illinois, fluorite acid, and in making opal glass and enamel. Some colorless crystals are used to make lenses and prisms.

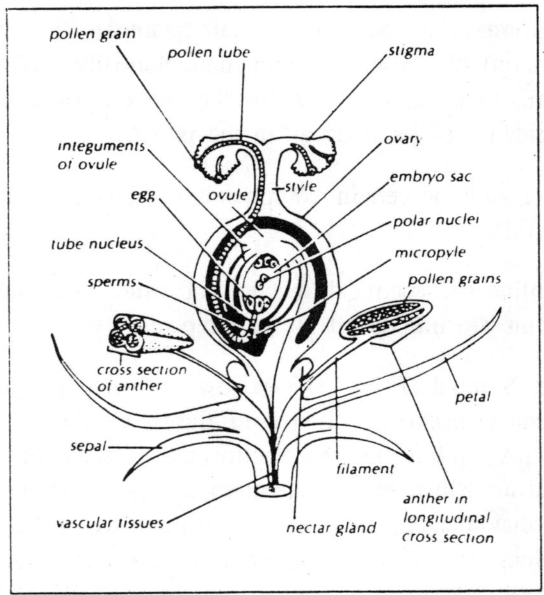

Fig. F-8. Flower: Longitudinal cross section of a flower at the time of fertilization, showing a pollen grain and pollen tube.

Fluorocarbon. Any of a class of compounds derived from hydrocarbons by replacing some or all of the hydrogen atoms by fluorine atoms.

Fluorspar (Fluorite). A mineral consisting of calcium fluoride, CaF_2. Fluorspar is the main natural source of fluorine.

Flux. 1. A measure of the strength of a field over an area.

2. A measure of a flow of energy or particles, usually expressed per unit area.

3. A substance that can react with an inorganic solid to produce a compound with lower melting point, used in the smelting to metals to form gangue into a molten slag.

4. A substance used in soldering to keep the surfaces clean and free from oxide.

Fluxmeter. An instrument for determining magnetic flux. One type consists of a ballistic galvanometer used in conjunction with a small coil (*search coil*) which is turned in the position investigated. Another type of fluxmeter depends on the voltage produced by the Hall effect in the field. Fluxmeters are sometimes called *gaussmeters.*

Fly. Name for a variety of winged Insects, but properly restricted to members of the order Diptera. True flies have sucking and piercing mouthparts and, in most species, one pair of wings. All flies undergo complete Metamorphosis. Most are harmful as disease carriers or crop destroyers. The order includes the Fruit Fly, Gnat, Mosquito, and Tsetse Fly.

Flying Fish. Torpedo-shaped Fish (family Exocoetidae) of warm seas with well-developed pectoral fins that can be held rigid and used to glide short distances over the water. Specially adapted tail fins, vibrated in taxiing along the water surface to gain momentum, and (in some species) enlarged pelvic fins also help in "flight". The largest species is the 18-in. (45 cm) California flying fish.

Flying Squirrel. Nocturnal Squirrel adapted for gliding; flying squirrels do not actually fly. Most are found in Asia, but there are two North American species (genus (*Glaucomys*) which together are found in forests over much of the continent. The gliding mechanism is a fold of skin extending along each side of the body. When the animal extends its limbs in leaping, the flaps stretch out taut like a parachute.

F-number. The ratio of the focal length of a lens to its diameter. The F-number is used in photography, different f-numbers of camera lenses being obtained by use of a variable diaphragm to change the effective lens diameter. Use of higher f-numbers decreases the image brightness and necessitates a longer exposure time. The f-number of a lens is sometimes referred to as its *speed.*

Foam. A dispersion of small bubbles of gas in a liquid or solid. The gas is usually air and the dispersion medium is usually a liquid. Foams in liquids are stabilized by the addition of foaming agents, usually surfactants. Solid foams are produced by foaming the substance as a liquid and allowing it to set.

Focal Length. The distance from the pole of a mirror or centre of a lens to the focal point.

Focal Point. Either of two points on each side of a lens, mirror, or other optical system, such that a ray parallel to the axis converges to or appears to diverge from one of these points.

Focus. A point to which rays of light are converged by a lens, mirror, or other optical system.

Fog. Aggregation of water droplets just above the earth's surface, i.e., a cloud near the ground. A light or thin fog is usually called a mist. Fog may occur either when the moisture content of the air is increased beyond the saturation point or when the air is cooled below the Dew point. In either case, excess moisture condenses on microscopic dust particles (condensation nuclei) in the atmosphere.

Fold. In geology, bent or deformed arrangement of stratified rocks. Arches, or upfolds, are known as anticlines; depressions, or downfolds, are called synclines. An imaginary line drawn along the crest of an anticline or the trough of a syncline is its axis; the two sides curving away from the axis are the limbs. The complex causes of folding involve large-scale crustal movements.

Food preservation. Because most food remain edible for only a limited time, food preservation has been practiced since the remotest times. Early products of conservation were cheese, butter, wine, bacon, pemmican, raisins, and parched grains. Advances in the methods used came with scientific investigations of the microorganisms that cause food spoilage, especially in the work of Louis Pasteur. Basic methods of modern food preservation include dehydration, freeze-drying, heating (e.g., Canning and Pasteurization), refrigeration (both freezing and chilling), hermetic sealing to remove air, and atomic radiation to destroy microorganisms. Such preserving agents as salt, vinegar, sugar, smoke, and alcohol are also used, often in combination. Other common preservatives are chemical agents.

Foot. An imperial unit of length equal to one third of a yard. It is equal to one third of a yard. It is equivalent to 0.3048 metres.

Force. Symbol: *F* The agency that can cause a body to alter its state of rest or of uniform motion. The unit of force is the newton.

Force. In physics, a quantity that produces a change in the size or shape or the Motion of a body. Commonly experienced as a "push" or "pull", force is a vector quantity, having both magnitude and direction. Four basic types of force are known in nature. The gravitational force and the electromagnetic force both have an infinite range. The strong nuclear force, or strong interaction, is a short-range force holding the atomic nucleus together, and the weak nuclear force, or weak interaction, is a short-range force associated with radioactivity and particle decay. In the Metric System forces are measured in such units as the dyne (cgs system) and the newton (mks system), which cause accelerations of, respectively, 1 cm/sec^2 on a 1-gram mass and 1 m/sec^2 on a 1-kg mass. In English Units of Measurement the Pound (lb) is used. A 1-lb force equals 444,823 dynes; 1 dyne equals 10^{-5} newtons.

Force Convection. *See* Convection.

Ford, Henry, 1863-1947. American industrialist, pioneer automobile manufacturer; b. Dearborn, Mich. While working as a machinist and engineer with the Edison Co., in his spare time he built (1892) his first automobile. In 1903 he organized the Ford Motor Co. By cutting production costs, controlling raw materials and distribution, adapting the assembly line to automobiles, and featuring an in expensive, standardized car, Ford became the largest automobile producer in the world. In 1908 he designed the Model T; over 15 million cars were sold before the model was discontinued (1928), and a new design, the Model A, was created to meet growing competition. In 1914 Ford created a sensation by paying his workers five dollars for an eight-hour day, considerably above the average, and by beginning a profit-sharing plan that would distribute up to $30 million annually among his employees. He stubbornly resisted union organization until 1941. He retired in 1945. His numerous philanthropies included the Ford Foundation.

Forestry. The management of forests for Wood, water, wild-life, forage, and recreation. Due to wood's economic importance, forestry has been chiefly concerned with timber management, especially reforestation, maintenance of extant forests, and fire control. The U.S. Forest Service manages the National Forest system (federally owned forest reserves), emphasizing sustained yield (achieved by balancing reforestation and logging) and multiple use.

Form. *See* Crystal form.

Formaldehyde (Methanal). A colorless poisonous gas, HCHO, with a pungent odour. It is a water-soluble aldehyde forming a solution known as *formalin,* which is used as a preservative. Formaldehyde is manufactured by the catalytic oxidation of methanol or low-boiling petroleum gases and is used in making synthetic resins. M.pt. -92°C; b.pt. -19°C.

Formaldehyde or **Methanal.(HCHO)**. A colorless, flammable, poisonous gas with a suffocating odor. Pure gaseous formaldehyde is uncommon, because it readily polymerizes into solid paraformaldehyde. Formalin, a 40% by volume solution of formaldehyde in water, is used as an antiseptic, disinfectant, and preservative for biological specimens. Formaldehyde is also used to make Dyes, Plastics (e.g., Bakelite), and synthetic Resins.

Formalin. *See* Formaldehyde.

Formate. Any salt or ester of formic acid.

Formic acid (methanoic acid). A colourless liquid, HCOOH, with a pungent odour, made by treating sodium formate with sulphuric acid. It is a strongly reducing carboxylic acid, used in finishing textiles and manufacturing chemicals. M.pt. 8°C; b.pt. 101°C; r.d. 1.2.

Formula. A representative of a molecule using symbols for the atoms. The *molecular formula* of a compound indicates the numbers of atoms present, as in $C_2H_4O_2$ for acetic acid. Usually suchformulas are written as combinations of functional groups, CH_3COOH. The *empirical formula* indicates the relative numbers of atoms and thus gives the simplest possible structure. Acetic acid has an empirical formula CH_2O. The *structural formula* shows the relative arrangement of the atoms in space.

Formyl Group. The monovalent group HCO^-.

Fortin Barometer. A type of mercury barometer. [After Nicolas Fortin (1750-1831), French instrument maker.]

Fortran. A computer programming language used particularly in scientific work.

Fossil. Remains or imprint of a plant or animal preserved from prehistoric times by natural methods and found mainly in sedimentary rock, asphalt, and coal. Fossilization of skeletal structures or other hand parts is most common. Conditions for fossilization include quick burial in an originally moist sediment or other material that prevents both weathering and decay. Shells and bones embedded in hardened sediment can be dissolved by water, leaving a natural mold. Sometimes these molds are filled with mineral deposits, forming natural casts. Fossil footprints and trails and coprolites, or fossil excrement, can reveal much about the lives and feeding habits of ancient animals. Entire animals of the late Pleistocene have been found, notably in Siberia. In California, the La Brea tar pits have yielded many skeletons. The study of fossils is paleontology.

Foucault Pendulum. A heavy metal ball suspended on a long wire, used to demonstrate the rotation of the earth by the fact that the plane of its motion rotates slowly. [After Jean bernard Foucault (1819-68), French physicist.]

Fourier, Jean Baptiste Joseph, Baron. 1768-1830, French mathematician and physicist, noted for his researches on heat diffusion and numerical equations. He originated the Fourier series, which allowed discontinuous functions to be represented by a trigonometric series.

Fourier Series. A series of form $a_o + a_1 \cos \omega t + b_1 \sin \omega t + a_2 \cos 2\omega t + b_2 \sin 2\omega t + ...$ Any periodic function f(t) can be expressed in this form, as a sum of harmonic terms. [After Jean Baptiste Fourier (1763-1830). French mathematician].

Fox. Carnivorous Mammal of the Dog family, found in much of the Northern Hemisphere. It has a pointed face, thick fur, and bushy tail. Most fox species belong to the red fox group (genus *Vulpes*). The common red fox (*V.vulpes*) of North America inhabits areas of forest mixed with open country from the Arctic to the S U.S.

Foxhound. Sturdy, medium-sized Hound. The *English Fox-hound,* shoulder height, 21-25 in. (53.3-63.5 cm); weight, 60-70 lb (27.2-31.8 kg), has a short, glossy coat that is usually black, tan, and white, probably developed from 14th-cent. French hounds, it was first used in packs to hunt foxes in the 17th cent. The slightly smaller American foxhound was developed over 300 years ago.

Fovea Centralis. *See* Eye.

FPS units. A system of units in which the base units are the foot, the pound, and the second.

Fraction. A quantity expressed in the form of one quantity divided by another. The dividend in a fraction is the *numerator* and the divisor is the *denominator:* in 7/8, 7 is the numerator and 8 the denominator. If both the numerator and denominator are integers the fraction is a *simple fraction.* Simple fractions are also called common *fractions* or *vulgar fractions.* A fraction in which both denominator and numerator are also fractions, as in (3/4)/(6/7), is a *complex fraction.* If the magnitude of the numerator is less than the denominator, then the expression is a *proper fraction*, otherwise it is an *improper fraction.* Thus, 3/7 is a proper fraction whereas 7/5 is improper.

Fraction. In arithmetic, an expression representing a part, or several equal parts, of a unit. In writing a fraction, e.g., 2/5 or 2/5, the number after or below the bar, called the denominator, indicates the total number of equal parts into which the unit has been divided; the number before or above the bar, called the numerator, indicates how many of these parts are being considered. The present notation for fractions is of Hindu origin, but some types of fractions were used by the Egyptians before 1600 B.C. The Decimal System uses another way to represent fractions, e.g., 0.75 for 3/4, often simplifying computation.

Fractional Crystallization: A technique for separating mixtures by utilizing differences in solubility or freezing point of the components. Typically, a solid mixture can be separated by dissolving it in a suitable solvent and lowering the temperature so that one component crystallizes out while the other remains in solution.

Fractional Distillation. The technique of separating liquid mixtures by distillation. The term is often used for distillation of a mixture in a container to which is attached a long vertical column filled with loose material, such as glass beads. As the vapour rises in the column, it condenses in the upper parts and the liquid runs back down the column (the *fractionating column*) until it revaporizes. Eventually, the more volatile components tend to collect in the upper part of the column with less volatile components lower down. This technique enables mixtures of liquids with similar boiling points to be separated efficiently. It is extensively used in the refining petroleum.

Frame of reference. A particular set of coordinates used as a reference system for making physical measurements.

Francium. Symbol : Fr. A radioactive metallic element. No stable isotope exists; the most stable, ^{223}Fr, has a half-life of 22 minutes. It is one of the alkali metals. A.N. 87; valency 1.

Francium (Fr). Radioactive element, discovered in 1939 by Marguerite Perey as a disintegration product of actinium. Some of the 21 Known isotopes of this rare Alkali Metal are prepared by bombarding thorium with protons, deuterons, or alpha particles.

Franklin, Benjamin. 1706-90, American statesman, printer, scientist, and writer; b. Boston. He went (1723) to Philadelphia as a printer and there won attention for his wit and commonsense philosophy, especially as expressed in *Poor Richard's Almanack* (pub. 1732-57). He helped establish (1751) an academy that became the present Univ. of Pennsylvania, and he served as deputy postmaster general of the colonies (1753-74). His famous experiment with a kite in a thunderstorm proved the presence of the electricity in lightning. Franklin proposed a plan of union for the colonies at the Albany Congress (1754) and was agent for several colonies in England. Returning (1775) to America during the Revolutionary unrest, he helped draft the Declaration of Independence, which he signed. During the American Revolution he was a successful American agent in France and was appointed (1781) a commissioner to negotiate the peace with Britain. Franklin's last great public service was his attendance at the Federal Constitutional Convention (1787). His autobiography is well known. Benjamin's illegitimate son, William Franklin, c. 1730-1813, was the last royal governor of New Jersey and a Loyalist in the American Revolution.

Frasch Process. A process for obtaining sulphur from underground deposits using a set of three concentric pipes sunk into the ground. Superheated steam is forced down the outer pipe and compressed air is forced down the centre to drive molten sulphur up to the surface.

Fraunhofer Lines. Dark absorption lines in the spectrum of radiation from the sun. They are caused by absorption of certain frequencies of light in the cooler outer regions of the sun's atmosphere. [After Joseph Fraunhofer (1787-1826), German instrument maker.]

Free Energy. A thermodynamic function used to measure the ability of a system to perform work. A change in free energy is equal to the work done.

Free Fall. In physics, the state of a body moving solely under the influence of gravitational forces. A body falling freely toward the surface of the earth undergoes an acceleration equal to 32 ft/sec^2 ($9.8\ m/sec^2$).

Free Radical. An atom or group of atoms that has an independent existence without all its valencies being satisfied. Thus ethane, which contains two methyl groups, can be split into two electrically neutral methyl radicals (homolytic fission). Radicals are usually written with a dot to represent their impaired electron, CH_3. Atoms produced by breaking diatomic molecules are also regarded as free radicals. Because of their unsatisfied valency species of this type are highly reactive and usually have only a transient existence. They may attack other molecules to abstract an atom and yield another radical. Two radicals can also recombine to give a neutral molecule.

Freeze drying. Very rapid freezing and drying of a substance in a high vacuum system, used to preserve foodstuffs, etc.

Freezing. The change of a substance from the liquid state to the solid state as a result of lowering the temperature. The temperature at which this occurs is the *freezing point* of the substance. It is the temperature at which the solid and liquid phases are in equilibrium at a specified (usually standard) pressure. The freezing point is equal to the melting point.

Freezing Mixture. Two or more substances mixed to give a low temperature. Simple examples of freezing mixtures are ice and sodium chloride (-20°C) and ice and calcium chloride (-40°C).

F-region. *See* Ionosphere.

Frenkel Defect. *See* Interstitial.

Freon. ® Any of a group of fluorocarbon products that are clear colourless liquids with an ethereal odour. All are chemically inert and of low toxicity. They are used as refrigerants, aerosol propellants, cleaning fluids, and solvents.

Freon. Trade name for any of a special class of chemical compounds used as refrigerants, aerosol propellants, and solvents. Freons are Hydrocarbon derivatives that contain fluorine and often chlorine and bromine as well. They are generally colorless, odorless, nontoxic, noncorrosive, and nonflammable. Though usually unreactive, freons are now suspected to undergo reactions in the upper atmosphere that may damage the earth's Ozone layer. The most commonly used is Freon-12, or dichlorodifluoromethane (CCl_2F_2).

Frequency. Symbol: f or v. The number of complete cycles of a periodic process occurring in unit time. The frequency of a wave is equal to c/λ, where c is the velocity and λ the wavelength. Frequency is measured in hertz.

Frequency Modulation. *See* Modulation.

Fresnel. A unit of frequency equal to 10^{12} hertz. [After Augustin Jean Fresnel (1788-1827), French physicist.]

Fresnel Lens. A type of lens having a convex surface and a stepped surface used in headlights, searchlights, etc.

Fig. F-9. Fresnel lens.

Fresnel's biprism. An experiment to demonstrate interference of light by placing a biprism of narrow angle in front of a monochromatic light source. The two images produced by the prism act as coherent sources of equal intensity and produce interference fringes.

Friction. A force that tends to oppose the relative motion of two surfaces in contact. The *coefficient of friction* (symbol: μ) is the ratio of the normal force of reaction between the surfaces to the frictional force. In fact the frictional force for a moving body is less than that opposing the movement of a static body, leading to two coefficients: one of *sliding friction* and the other of *static friction.*

Friction. Resistance offered to the movement of one body past another body with which it is in contact. The amount of friction depends on the nature of the contact surfaces and on the magnitude of the force pressing the two bodies together, but not on the surface area of the contact surface. The coefficient of friction is the ratio of the force necessary to move one body horizontally over another at a constant speed to the weight of the body. Fluid friction, observed in the flow of liquids and gases, is minimized in airplanes by a modern, streamlined design.

Friedel-Crafts reaction. A type of chemical reaction in which an alkyl or acyl group is substituted into a benzene ring by reaction of an alkyl or acyl halide. The two types of reaction can be illustrated by simple examples.

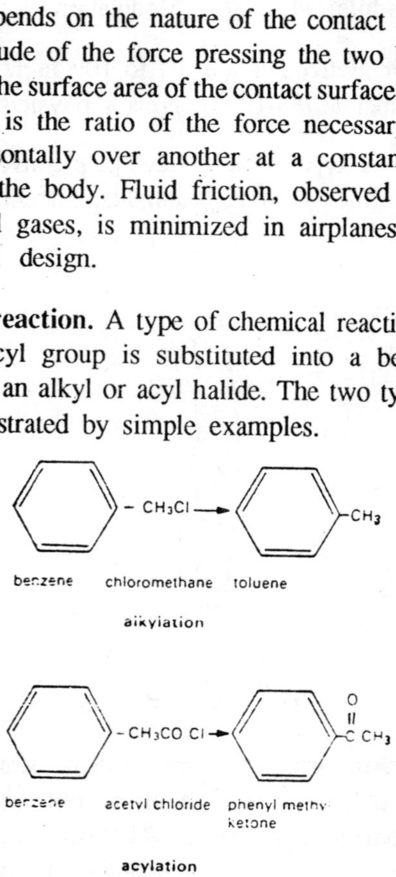

– CH_3Cl → –CH_3

benzene chloromethane toluene

alkylation

– CH_3CO Cl → –C($=$O) CH_3

benzene acetyl chloride phenyl methyl ketone

acylation

Fig. F-10. Friedel-Crafts reactions.

All Friedel-Crafts reactions are catalysed by aluminium

chloride, boron trifluoride, tin IV chloride, and similar compounds. [After Charles Friedel (1832-99), French chemist, and James M. Crafts (1839-1917), American chemist].

Fringes. Alternate light and dark bands produced by interference or diffraction.

Frisch, Karl von. 1886-, Austrian zoologist. For his pioneering work in comparative behaviorial psychology, particularly his studies of the complex communication between insects, he shared the 1973 Nobel Prize in physiology or medicine with Konrad Lorenz and Nikolaas Tinbergen. An important implication of his work is that behavioral continuity exists between animal communication and human language.

Frog. Tailless, freshwater Amphibian (order Anura) found worldwide. Some frogs require moisture and are highly aquatic; others, such as Toads, are more terrestrial. Frogs have bulging eyes, short, neckless bodies, long, muscular hind legs for jumping, webbed feet for swimming, and smooth skin, usually green or brown. They capture insects and other food with a sticky, forked tongue. Most frogs lay eggs in early spring, and by the end of summer Metamorphosis to a four-legged adult is complete. The so-called true frog belongs to the Ranidae family, which includes the bull-frog (*Rana catesbeiana*) and other North American frogs.

Front. In meteorology, the boundary between adjacent air masses. If a cold air mass pushes a warm air mass ahead of it, a cold front exists; a warm front is the reverse situation. A stationary front is one in which neither air mass pushes against the other. An occluded front exists when a cold front catches up to a warm front; the warm air mass lying between the two cold air masses is thus forced to rise.

Frost or **Hoarfrost.** Ice formed by the condensation of atmospheric water vapor on a surface whose temperature is below 32°F (0°C). In the formation of frost, water vapor is changed directly to a solid. Frost appears as a light, feathery deposit of ice, often in a delicate pattern.

Frostbite or **Chilblains.** Tissue injury caused by exposure to cold, usually affecting the hands, feet, ears, or nose. Extreme cold causes the small blood vessels in the extremities to constrict, resulting in slowed blood circulation and stagnation, which deprive tissues of nutrients. The condition is aggravated by inactivity and dampness. Severe, untreated frostbite may result in Gangrene.

Froth Flotation. Any of several processes in which materials, notably minerals, are separated by agitation of a pulverized mixture of the material with water oil, and chemicals that cause differential wetting of the suspended particles, the unwetted particles being carried by air bubbles to the surface for collection.

Fructose (fruit sugar). A white crystalline simple sugar, $C_6H_{12}O_6$, that occurs naturally in fruits and honey. M.pt. 103°C; r.d. 1.6.

Fructose, levulose, or **fruit sugar.** Simple Sugar found in honey and fruit. Sweeter than Sucrose, fructose in a carbohydrate with the same fomula as Glucose, but with a different structure. An equimolar mixture of fructose and glucose, called invert sugar, is obtained by the breakdown of sucrose and the is major component of honey.

Fruit. Matured ovary of a Flower, containing the seed. After Fertilization takes place and the embryo plantlet has begun to develop, the surrounding ovule becomes a seed, and the ovary wall around the ovule (pericarp) becomes the fruit. Fruits are classified into four types: simple, aggregate, multiple, and accessory. Simple fruits are either fleshy or dry. Fleshy fruits are either berries (the entire pericarp is fleshy, e.g., the blueberry, tomato, and banana) or drupes (the inner layer of the pericarp becomes a pit or stone around the seed, e.g., the

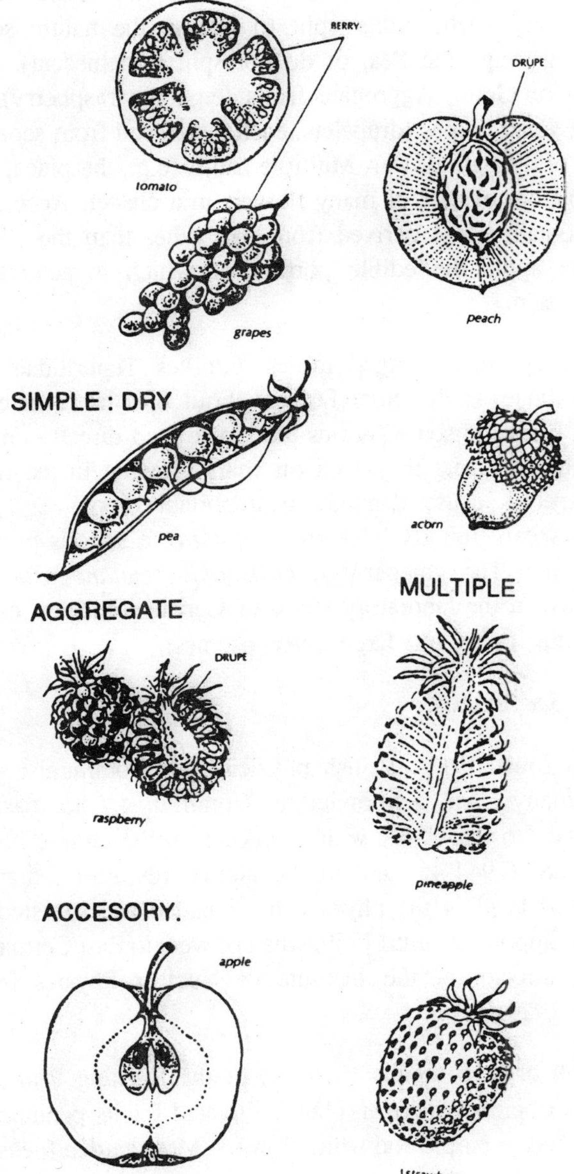

Fig. F-11. Types of fruit.

peach and walnut). Dry fruits are divided into those whose hard, papery shells either split to release the nature seeds (dehiscent), e.g., the Pea, or do not split (indehiscent). e.g., the Nut or Grain. Aggregate fruits (e.g., the raspberry) are mass of small drupes (drupelets, each developed from separate ovaries of a single flower. Multiple fruits. (e.g., the pineapple) develop from ovaries of many flowers in a cluster. Accessory fruits contain tissue derived from part other than the ovary, e.g., the apple the edible portion of which is actually a swollen stem.

Fruit fly. Very small winged insect (families Tephritidae and Drosophilidae) in the order Diptera. Fruit flies lay their eggs in plant tissue. In some species the larvae feed directly on the fruit pulp; in others they feed on years growing in the fruit. Some species cause damage to important crops; e.g., the Mediterranean fruit fly (*Ceratitis capitata),* a serious pest of citrus fruits. The vinegar fly, *Drosophila melanogaster,* is often used in the laboratory study of Genetics because of its ten-day life cycle and large chromosomes.

Fruit sugar. *See* fructose.

Fuchs, Klaus Emil. 1911-, British physicist and Communist spy; b. Germany. An acknowledge Communist, he passed information to the USSR while working on the atomic bomb in the U.S. (1944-45) and in the atomic research center at Harwell as head of the physics division. He was arrested in 1950 and imprisoned until 1959, when he went to East Germany. He was director of the Institute of Nuclear Physics from 1959 to 1979.

Fuchsia. Shrub or herb (genus *Fuchsia)* of the Evening Primrose family, a tropical American plant cultivated for its pendulous, brilliant, red-to-purple and white flowers. Most garden fuchsias are hybrids.

Fuel cell. A type of electric cell in which electrial energy is produced directly by electrochemical reactions involving substances that are continuously added to the cell. In the simplest type hydrogen is the fuel and oxygen is the oxidizer. Each gas is fed to a porous metal electrode, usually of nickel or platinum, in an electrolyte of aqueous potassium hydroxide or a molten salt. Electron transfer occurs at the electrodes, the hydrogen electrode being the cathode, and within the cell hydrogen and oxygen combine to produce water.

Fuel cell. Electric Cell in which the chemical energy from the oxidation of a gas fuel is converted directly to electrical energy in a continuous process. In the hydrogen and oxygen fuel cell, hydrogen and oxygen gas are bubbled into separate compartments connected by a porous disk through which an Electrolyte, such as aqueous potassium hydroxide (KOH), can pass. Inert graphite electrodes, mixed with a catalyst such as platinum, are dipped into each compartment. When the two electrodes are electrically connected, an Oxidation and Reduction reaction takes place in the cell; hydrogen gas is oxidized to form water at the anode; electrons are liberated in this process and flow through the external circuit to the cathode, where the electrons combine with the oxygen gas and reduce it. Fuel cells have been used to generate electricity in spacecraft.

Fulcrum. The pivot about which a lever turns.

Fuler's earth. A type of clay used as a catalyst and absorbent.

Full radiator. *See* black body.

Full wave rectification. Rectification of an alternating current or voltage in such a way that alternate half cycles are reversed in direction, so that the signal repeatedly increases to a maximum and decreases to zero. *Compare* half-wave rectification.

Fulminate. *See* cyanic acid.

Fulminic acid. *See* cyanic acid.

Fumaric acid (*trans*-butenedioic acid). A colourless crystalline carboxylic acid with a sharp fruity taste, HOOCCH : CHCOOH, used in the manufacture of synthetic resins. Fumaric and maleic acids are geometrical isomers. Sublimes at 200°C; r.d. 1.64.

Fuming nitric acid. *See* nitric acid.

Fuming sulphuric acid.*See* sulphuric acid.

Function. A mathematical expression involving variables.

Function. In mathematics, a relation that assigns to each member x of some set X (the domain) a unique member y of some set Y (the range); y is said to be a function of x, usually denoted $f(x)$ (read "f of x"). In the equation $y = f(x)$, x is called the independent variable and y the dependent variable. Although a function f assigns a unique y to each x, several x's may yield the same y; e.g., if $y = f(x) = x^2$ (where x is a number), then $f(2) = f(-2) = 4$. If this never occurs, then f is called a one-to-one, or injective, function.

Functional group. In organic chemistry, group of atoms within a molecule that is responsible for certain properties of the molecule and reactions in which it takes part.

Functional group. *See* group.

Fundamental. The component with the lowest frequency in a complex wave.

Fungi. Division of simple plants that lack Chlorophyll, true stems, roots, and levels. Unlike Algae, fungi cannot photosynthesize, and live as parasites or Saprophytes. The division comprises the Slime Molds and true fungi. True fungi are multicellular (with the exception of Yeasts); the body of most true fungi consists of slender cottony filaments, or hyphae. All fungi are capable of asexual Reproduction by cell division, budding, fragmentation, or Spores. Those that reproduce sexually alternate a sexual generation (gametophyte) with a spore-producing one. The four classes of true fungi are the algaelike fungi e.g., black bread Mold and downy Mildew), sac fungi (e.g., yeasts, powdery mildews, Truffles, and blue and·green molds such as *Penicillium),* basidium fungi (e.g., Mushrooms and puffballs) and imperfect fungi (e.g., species that cause athlete's foot and ringworm). Fungi help decompose organic matter (important in soil renewal); are valuable as a source of Antibiotics, vitamins, and various industrial chemicals; and for their role in Fermentation, e.g., in bread and alcoholic beverage production.

Fungicide. Substance used to prevent or destroy Fungi. Made from sulfur or copper compounds, organic salts or iron, zinc, and mercury, and other chemicals, fungicides are used on seeds, soil, wood, and fabrics, and for human fungus diseases as well as fungus diseases of plants.

Funk, Casimir. (foongk), 1884-1967, American biochemist; b. Warsaw, Poland; came to U.S., 1915. Credited with discovering Vitamins, he stirred public interest with his 1912 paper on vitamin-deficiency diseases. He coined the term *Vitamine* and later posited the existence of four such materials (B_1, B_2, C, and D). He contributed to knowledge of the hormones of the pituitary and sex glands and stressed the importance of balance between hormones and vitamins.

Fur. Hairy covering of an animal, especially that with thick soft, close-growing underfur next to the skin and a coarser protective layer of guard hair above it. The term includes sheepskins with their hair, and curled pelts, e.g., Persian lamb. Used for clothing since prehistoric times, furs were traditionally luxury goods in the more populous, temperate regions of the world, where wild animals are scarce. The most prized furs include sable, mink, and chinchilla, but many others, e.g., beaver, fox, rabbit , and ocelot, are also valued, and breeding and hunting industries are important in North America and Asia. Following the exploitation (17th-19th cent.) of N American and Asian wilderness areas, depleting the sea otter and threatening the fur seal, conservationists and humane groups have protested trapping for fur. In 1969 the U.S. government passed the Endangered Species act, banning important or sale of pelts of such animals as the polar bear, jaguar, and tiger. Since World War II synthetic furs have become increasingly popular because of their realistic appearance, the relatively low cost, and public awareness of endangered species.

Fur. Calcium carbonate deposit on the inside of kettles, boilers, hot-water pipes, etc. It is precipitated by the decomposition of calcium bicarbonate present in hard water.

Fig. F-12. Furan.

Furan. A colourless liquid. C_4H_4O. M.pt.-86°C; b.pt. 31.4°C; r.d. 0.94.

Furanose. *See* sugar.

Furniture. Such movables as chairs, tables, beds, chests, and cabinets; the term may also be used for draperies, rugs, mirrors, and lamps. Since ancient times furniture has been made of many materials, with styles and techniques showing great variety, but in the past most pieces were fashioned of wood and decorated by Inlaying, painting or gilding, carving veneering, and marquetry. Ancient Oriental furniture displays carving and inlay on ebony and teak. Egyptian pieces 6,000 years old have animal carvings and gold and ivory inlay. The Greeks used low couches and curved chairs. The Romans adopted Greek and Etruscan forms. Heavily carved Gothic furniture reflected architectural styles. Renassisance pieces were richly decorated. Peasant furniture was solid, painted or carved, and show to change in style. Provincial pieces were simple and of native woods. Period styles, such as the Louis Period styles, Directoire Style, and Empire Style, developed in cultural centers. English period styles include Elizabethan, Jacobean, Queen Anne, and Georgian, and were represented by such designers as Chippendale, Hepplewhite, and Sheraton. Early American cabinetmakers adapted English system in utilitarian form, using such native woods as pine, maple, and cherry; later, Phyfe and others added walnut and mahogany. In the 19th cent. mass production of furniture began. Early in the 20th cent. Art Nouveau influenced furniture design. Later, such architects as Rietveld, Mies Van Der Rohe, and Eero Saarinen developed functional pieces using modern materials. After World War II, elegant, simple Scandinavian furniture of fine woods became popular worldwide. Trends in the late 1970s to early 1980s included the austere, minimal look of built-ins and a return to the opulent, padded lock of Art Deco.

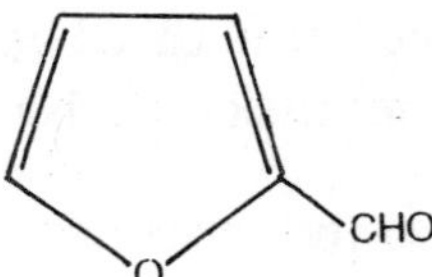

Fig. F-13. Furfural.

Furfural. A colourless liquid cyclic aldehyde, $C_5H_4O_2$ prepared from cereal straw or bran and used as a solvent and in the manufacture of chemicals. M.pt. -36°C; b.pt. 162°C; r.d. 1.2.

Fuse, electric. Safety device used to protect an Electric Circuit against an excessive current. A fuse consists of a low melting-point alloy strip enclosed in a suitable housing and connected in series with the circuit that it protects. Because of its electrical Resistance, the alloy strip is heated by electric curent; if the current exceeds the safe value for which the fuse was designed, the strip melts, opening the circuit and stopping the current.

Fused. Denoting a solid that has been melted and resolidified to a solid mass.

Fused ring. A ring of atoms in a molecule attached to another ring in such a way that two atoms are common to both rings. Naphthalene is a simple example of a fused-ring compound.

Fusel oil. Oily, colourless liquid with a disagreeable odor and taste. A mixture of Alcohols and fatty acids formed during the alcoholic Fermentation of Starch, fusel oil may occur as an unwholesome impurity imperfectly distilled liquor. It is used as a solvent in making lacquers and enamels.

Fusion. 1. Melting.

2. A nuclear reaction in which two light nuclei join to form a heavier nucleus. An example is the formation of helium from hydrogen: ${}^{2}_{1}H + {}^{3}_{1}H = {}^{4}_{2}He + n$. Like all fusion reactions it is accompanied by evolution of large amounts of energy. The nuclei must first have a high kinetic energy to overcome their electrostatic repulsion and this happens at high temperatures. Fusion reactions are often called *thermonuclear reactions. See* nuclear weapon, fission, carbon cycle.

Fusion reactor. A device for obtaining energy by nuclear fusion. In general, a fusion reactor has a configuration of magnetic fields to hold the plasma, which is produced initially by a pulse of high current through the gas. The main problems are in maintaining a stable plasma and in extracting the energy used So far fusion reactors are only in the experimental stage.

G

Gadolinium. Symbol: Gd. A silvery white malleable ductile metallic element belonging to the lanthanide series. A.N. 64; A.W. 157.25; m.pt. 1311°C; b. pt. 3233°C; r.d. 7.9; valency 3.

Gadolinium (Gd). Metallic element, extracted in oxide form by J.C.G. de Marignac in 1880. This silver-white, malleable, ductile, lustrous Rare-earth Metal is found in godolinite, Monazite, and bastnasite. It is paramagnetic at room temperature but becomes strongly ferromagnetic when cooled.

Gain. The increase in signal produced by an amplifier in an electronic circuit. The gain is usually expressed in decibels as the ratio of the output power to the input power.

Galaxy. Any of the innumerable aggregations of stars that, together with gas, dust, and other material, make up the universe. Each galaxy contains, on average, about 100 000 million stars. The vast majority of galaxies are regular systems, either *spiral galaxies*, like our Milky-Way System, or *elliptical galaxies*. The distance between one galaxy and another runs to millions of light-years but galaxies usually form groupings. Our own system has two satellite galaxies, the two Magellanic Clouds visible in the southern hemisphere.

Galaxy. Large aggregation of gas, dust, and typically billions of stars. It is held together by the gravitational attraction between its constituent parts, and its rotational motion prevents it from collapsing on itself. A typical spiral galaxy is shaped like a flat disk, about 100,000 light-years in diameter, with a central bulge, or nucleus, containing old stars; winding through the disk are the characteristic spiral arms of dust, gas, and young stars. An elliptical galaxy, lacking spiral arms entirely and containing little or no gas and dust, resembles the nucleus of a spiral galaxy. A small minority of galaxies are classified as irregular, i.e., showing no definite symmetry or nucleus. One theory suggests that irregulars evolve into spirals or ellipticals depending on the initial amount of rotational motion. Some galaxies radiate a large fraction of their energy in forms other than visible light, such as radio waves, X rays, and infrared and ultraviolet radiation; their optical counterparts may be faint or undertectable. Gravitation also holds clusters of galaxies together, the Local Group cluster includes the Milky Way (containing the sun and solar system) and the Andromeda Galaxy, both spirals, and the irregular Magellanic Clouds. See also Hubble's Law.

Galaxy (Milky-Way System). The spiral galaxy in which the sun is located. It is shaped like a Catherine wheel with a convex central nucleus. The Galaxy's diameter is 100 000 light-years and its thickness measured through the nucleus is about 20 000 light-years. It belongs to a local group of 20 or so galaxies including the one in Andromeda. *See also* galaxy.

Galena (lead glance). A grey mineral consisting of lead II sulphide, PbS. It is the principal ore of lead and also contains small amounts of silver.

Galilean Telescope. A simple terrestrial telescope with a convex objective lens and a concave eyepiece separated by the difference between the focal lengths of the two lenses. The system gives an erect image and is used in low-powered opera glasses and simple binoculars. [After Galileo Galilei (1564-1642) Italian scientist].

Gall, Franz Joseph, 1758-1828. Austrian anatomist Devoting most of his life to a minute study of the nervous system, he showed that the brain's while matter consists of nerve fibers; he also launched the doctrine of localization of various mental processes in parts of the brain. Gall founded the pseudoscience of phrenology.

Gall bladder. Small, pear-shaped sac attached to the Liver by the hepatic duct. The gall bladder stores and concentrates Bile, which is produced by the liver and functions in fat digestion. Fatty foods stimulate intestinal cells to produce a hormone that causes the gall bladder to contract, forcing the bile into the common bile duct and the intestine, where fats are absorbed. Components of bile sometimes crystallize in the gall bladder, forming gallstones.

Gallium. (Ga), metallic element, discovered spectroscopically by P.E. Lecoq de Boisbaudran in 1875. Solid gallium is blue-gray; the liquid form is silver. It is chemically similar to Aluminum. Gallium compounds are used in transistors, lasers, and diodes. See Element (table); Periodic Table.

Gallium. Symbol: Ga. A silvery-white metallic element present in small amounts in some ores, including bauxite. It is a low-melting substance and the liquid is more dense than the solid. Gallium arsenide is an important semiconductor. A.N. 31 A.W. 69.72; m.pt. 29.8 °C; b. pt. 2403°C; r.d. 5.9°C.

Gallon. *See Appendix.*

Galvanic cell. *See* primary cell.

Galvani, Luigi. 1737-98, Italian physician. A noted surgeon and researcher in comparative anatomy. Galvani concluded, from his observation that a frog's leg contracted when touched by two different metals in a moist environment, the animal tissues generate Electricity. Although he was eventually proved wrong by Alessandro Volta, the controversy stimulated research in electrotherapy and on electric currents. Many electrical terms are derived from Galvani's name.

Galvanized iron. Iron or steel coated with a thin layer of zinc to protect it from corrosion.

Galvanizing. The process of coating iron or steel with zinc to produce galvanized iron. Zinc is applied to the metal from a molten bath containing a flux.

Galvanizing. Process of coating a metal, usually iron or steel, with a protective coating of Zinc. Galvanized iron is usually prepared by dipping or electroplating. Pure iron, copper iron, and various steels are often galvanized, because the zinc coating resists oxidation and moisture.

Galvanometer. Instrument used to determine the presence, direction, and strength of electric current in a conductor. Galvanometers are based on the discovery (1819) by Hans Oersted that a magnetic needle is deflected by the presence of an electric current in a nearby conductor. When an electric current is passing through the conductor, the magnetic needle tends to turn at right angles to the conductor, so that is direction is parallel to the lines of induction around the conductor and its north pole points in the direction in which these lines of induction flow. In general, the extent to which the needle turns is dependent upon the current strength. In the modern d'Arsonval galvanometer, the magnet is fixed and the coil is movable. If a pointer is attached to the moving coil so that it passes over a suitably calibrated scale, the galvanometer can be used to quantitatively measure the current passing through it. A direct current (DC) Ammeter measures direct current by allowing a known percentage of the current to pass through a galvanometer. A DC Voltmeter measures direct voltage (see potential, Electric) between two points by allowing current to flow through a galvanometer connected in series with a high resistance. The current through the galvanometer is then proportional to the voltage (see Ohm's Law).

Galvanometer. Any of several instruments used for detecting or measuring small electric currents. Galvanometers are not usually calibrated in current units (*see* ammeter). The main types are the moving-coil and tangent galvanometers.

Games, theory of. Group of mathematical theories applying statistical logic to the choice of strategies in a game. A game consists of a set of rules governing a competitive situation in which two or more individuals or groups attempt to maximize their own winnings or minimize those of their opponents. Game theory, first developed by John Von Neumann, is applied to many fields, e.g., military problems and economics.

Gametophyte. Phase of a plant life cycle in which the egg (see Ovum) and Sperm (gametes) are produced. In many lower plants the gametophyte phase is the dominant life form; for example, the familiar Mosses are the gametophyte form of the plant. The union of egg and sperm gives rise to the sporophyte phase, in which spores are formed; spores, in turn, give rise to the gametophyte. In mosses the sporophyte is a capsule atop a slender stalk that grows out of the top of the gametophyte. The alternation between the gametophyte and sporophyte phase is known as alternation of generations.

Gamma iron. The allotropic form of iron stable between 906°C and 1403°C. It is nonmagnetic and has a face-centred cubic crystal structure.

Gamma radiation. Radiation emitted in one of the three types of natural Radioactivity. It is the most energetic form of Electromagnetic Radiation, with a very short wavelength of less than 10^{-10} m. Gamma rays are essentially very energetic X Rays emitted by excited nuclei. They often accompany alpha or beta particles, because a nucleus emitting those particles may be left in an excited (higher-energy) state. Gamma-ray sources are used in medicine for cancer treatment and for diagnostic purposes, and in industry for the inspection of castings and welds.

Gamma rays. Electromagnetic radiation with wavelengths in the range 10^{-10} — 10^{-13} metre. Gamma rays have shorter wavelengths than X-rays and higher photon energies. They are produced in some radioactive decay processes.

Gamow George, (gam'of), 1904-68. Russian-American theoretical physicist and author; b. Odessa; came to U.S., 1933. He is best known as an author who popularized abstract physical theories. Gamow devised (1928) a theory of radioactive decay and applied nuclear physics to problems of stellar evolution. He also proposed an important theory on the organization of genetic information in the living cell.

Gangrene. Local death of body tissue. Dry gangrene, the most common form, follows a disturbance of blood supply to the tissues, e.g., in Diabetes or destruction of tissue from injury. Moist gangrene results from an invasion of toxin-producing bacteria that destroy tissue. Treatment includes rest and Antibiotics; excision of the diseased area or, in advanced cases, amputation of the affected part may be necessary.

Gangue. The stony material occurring with a metal ore.

Gar. Cylindrical, mostly freshwater Fish (family Lepisosteidae) with hard, diamond-shaped scales, a long jaw, and long, sharp teeth, found in the U.S. and Central America. The largest species is the 9-ft (2.7-m) alligator gar *(Lepisosteus spatula)* of the Mississippi valley.

Gardenia. Evergreen shrub or tree (genus *Gardenia)* of the Madder family, native to the Old World tropics. Gardenias heavily fragrant and showy blossoms make them popular corsage and greenhouse plants. Most of the cultivated types are varieties of *G. jasminoides,* also called Cape jasmine but unrelated to true Jasmine.

Gas. A state of matter characterized by ease of flow, compressibility, and spontaneous expansion to fill any container in which the substance is placed. The intermolecular forces in gases are very small and the atoms and molecules show completely random behaviour. *See also* ideal gas, kinetic theory.

Gas carbon. A hard amorphous form of carbon produced as a deposit on the walls of the retort in the manufacture of coal gas.

Gas chromatography. *See* chromatography.

Gas constant. The constant R in the equation of state of an ideal gas (*see* gas laws). Also called the *universal gas constant*. R has the value 8.314 34 joules per kelvin per mole.

Gas-cooled reactor. *See* fission reactor.

Gas laws. Physical laws describing the behaviour of a gas under various conditions of volume (V), pressure (P), and absolute, or Kelvin, Temperature (T). Boyle's, or Mariotte's, gas law states that under constant temperature $PV = k_1$ Charles's, or Gay-Lussac's, law states that under constant pressure $V = k_2T$. A third law states that under constant volume $P = k_3T$. The constants K_1, k_2, *and* k_3 are dependent on the amount of gas present and, respectively, on the temperature, pressure, and volume of the gas. These three laws can be combined into a single law, or equation of state: $PV = kT$ or $Pv = RT$, in which v is the specific volume equal to V/n, n is the number of moles of the gas, k is a proportionality constant, and R is the universal gas constant, equal to 8.3149 X 10^3joules /kg-mole-degree in mks units. These laws are formulated for so-called ideal or perfect gases. Real gases are described more accurately by the van der Waals equations : $(P + a/v^2)$ $(v - b) = RT$, in which a and b are specific constants for each gas.

Gas Laws. Laws describing the behaviour of gases: the two gas laws are Boyle's law ($pV = C, T$ constant). The laws strictly apply only to ideal gases although real gases obey them at low pressure. The laws are combined in the expression $pV = nRT$, where n is the number of moles of gas and R is the gas constant. This is the equation of state of an ideal gas. Other equations are used for real gases, the best known being van der Waals equation.

Gas-liquid chromatography. *See* chromatography.

Gasohol. A gasoline extender made from a mixture of gasoline (90%) and ethyl alcohol (10%), often obtained by fermenting agricultural crops or crop wastes. Gasohol has higher octane, or "antiknock," properties than gasoline.

Gasoline. Light, volatile fuel oil, called petrol in Britain. A mixture of Hydrocarbons obtained in the fractional Distillation and "Cracking" of Petroleum, it is used as a fuel for Internal-Combustion Engines, for cooking, and as a solvent. The quality of gasoline used in engines is rated by Octane Number. To increase octane rating, lead additives were once widely used. Because of the possible health hazard of lead as an environmental pollutant and the harmful effect it has on pollution-control devices, however, a program was begun in the U.S. in the 1970s to change Automobile design and gasoline composition, so that lead additives could be gradually eliminated.

Gas thermometer. A thermometer consisting of a bulb containing gas attached to a pressure measuring device. Two types of gas thermometer are in use. In the constant-volume gas thermometer the volume is maintained at a constant value and the temperature is measured by the pressure of gas in the bulb. In the constant-pressure instrument the temperature is determined by the volume of gas, measured at constant pressure. The temperature is strictly proportional to the pressure or volume only for an ideal gas and corrections are applied in accurate work.

Gastropod. Mollusk of the class Gastropoda. It usually has a coiled or spiraled one-piece shell (univalve), although it may be reduced or absent. The head has sensory tentacles and a mouth with a rasplike tongue (radula). The ventral surface of the animal is modified into a large, flattened foot, which, along with other soft body parts, can be withdrawn into the shell; in most cases, the opening is covered by a plate (operculum). Most gastropods are marine, but there are forms that live in fresh water and on land. The class includes the Abalone, Conch, Snail, Slug, Whelk, Limpet Periwinkle, and Sea Slug.

Gate. 1. An Electronic circuit that gives an output signal only for certain combinations of two or more input signals. The types used in digital computers have two inputs and the signals can be either high or low. Gates can be designed to give a high output if both inputs are high, a high output if either input is low, etc.

2. The electrode that affects the current flow through the channel in a field-effect transistor.

Gatling, Richard Jordan. 1818-1903, American inventor; be Winton, N.C. He invented and manufactured agricultural implements, but is remembered as the creator of a rapidfiring gun, the precursor of the modern machine gun (see Small Arms). He offered it to the Union army, but it was not accepted for use until 1866, after the Civil War had ended.

Gatterman reaction. *See* Sandmeyer reaction.

Gauss. Symbol : G. A CGS—electromagnetic unit of magnetic flux density equal to a flux density of one maxwell per square centimetre. It is equivalent to 10^{-4} tesla. [After Karl Friedrich Gausss (1777-1855), German mathematician and physicist.]

Gauss, Carl Friedrich. (gous), 1777-1855, German mathematician, physicist, and astronomer; b. Johann Friederich Carl Gauss. Considered the greatest mathematician of his time and the equal of Archimedes and Isaac Newton, Gauss made many discoveries before he was twenty. His greatest work was in Number Theory; his *Disquisitiones Arithmeticae* (completed 1789; pub. 1801) is a masterpiece. Extremely careful and rigorous in his work, Gauss refused to publish any result without a complete proof. Consequently, many discoveries were not credited to him and were remade by others later, e.g., the work of Janos Bolyai and Nikolai Lobachevsky in Non-Euclidean Geometry, Augustin Cauchy in complex variable analysis, Carl Jacobi in elliptic functions, and Sir William Rowan Hamilton in quanternions. Gauss early discovered, independently of Adrien Legendre, the method of least squares. In 1801, when the asteroid Ceres was discovered by the Italian astronomer Giuseppe Piazzi, Gauss calculated its orbit on the basis of only a few accurate observations, and it was found the next year precisely where he had predicted. His *Theoria motus corporum celestium* (1809) treats the calculation of the orbits of planets and comets from observational data. From 1807 until his death Gauss was director of the astronomical observatory at Gottingen. Geodetic survey work done for the governments of Hanover and Denmark from 1821 led him to an interest in space curves and surfaces, as well as to the invention of the heliotrope, a device to measure distances by means of reflected sunlight, Gauss's collaboration with the German physicist Wilhelm Weber in research on electric and magnetic phenomena led to his invention, in 1833, of the electric telegraph.

Gaussian units. A system of CGS unit using both electrostatic and electromagnetic electrical units.

Gauss' law. The principle that the total outward flux of electric field strength E taken over any closed surface is the algebraic sum of the electric charges inside the surface divided by the electric constant.

$$\phi \, E ds = \Sigma Q/\varepsilon_0$$

The integral is over the whole surface. If EEo is replaced by D, the electric displacement, the integral is the electric flux (ψ), which equals the sum of the internal charges — this is an alternative statement of the law.

Gauss' law can be extended to other fields. In magnetism, the total outward magnetic flux (Φ) is zero.

Gaussmeter. An Instrument for measuring magnetic flux density.

Gay-Lussac, Joseph Louis. 1778-1850, French chemist and physicist. He and the French chemist Louis Jacques Thenard were the first to isolate Boron. Gay Lussac discovered (1802) independently that at constant pressure the volume or an enclosed gas is directly proportional to its temperature, a law known as Charles's law or as Gay-Lussac's law (see Gas Laws). He was the first to formulate (c. 1808) the law of combining volumes, which states that gases combine by volume in simple multiple proportion.

Gay-Lussac's law. The principle that gases react in volumes that bear a simple ratio to each other and to the volumes of the products if they are gases, all volumes being taken at the same temperature and pressure. Gay-Lussac's law is explained by Avogadro's hypothesis (equal volumes of different gases contain the same number of atoms).

Gegenschien (counter-glow). A patch of light seen in the night sky in a position opposite to that of the sun. The phenomenon is rare: like zodiacal light it is caused by the scattering of sunlight by meteor particles.

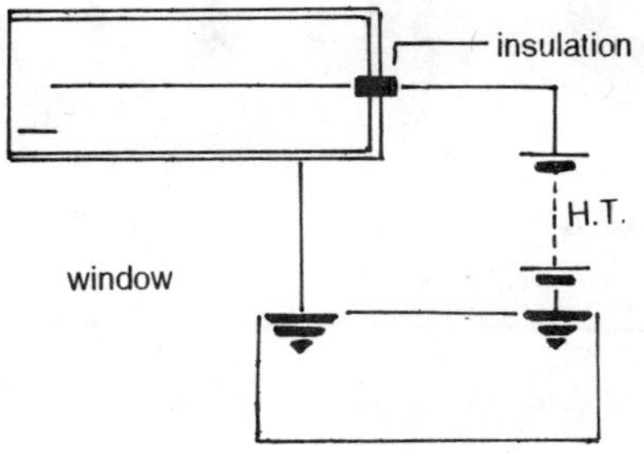

Fig. G-1. Geiger counter.

Geiger counter. A counter for ionizing radiation, consisting of a cylindrical metallic container with the thin metal window in one end and with a thin wire along the axis. The counter contains a low pressure of gas and high voltage (up to 1000 volts) is applied between the outer metal case (the cathode) and the axial wire (the anode). A particle or photon passing through the tube produces ionization. Electrons are accelerated towards the anode and ions move towards the cathode. On their way they produce further ionization and an avalanche occurs. The gas in the tube contains halogen compounds or inert gases and the discharge is quickly quenched. The pulse of current is amplified and registered on a counting circuit. [After Hans Geiger (1882-1945), German physicist.]

Gazelle. delicate, graceful Antelope (genus *Gazella)* inhabiting arid, open country, usually in Africa. Standing 2 to 3 ft (60 to 90 cm) high at the shoulder, the fawn-colored gazelle often has heavily ringed horns that curve back and inward. Gazelles are powerful jumpers and very swift. Closely related is the redish-brown impala *(Aepyceros melampus),* the most powerful jumper of all antelopes, able to leap 10 ft (3 m) into the air and travel 30 ft (9m) in a single bound.

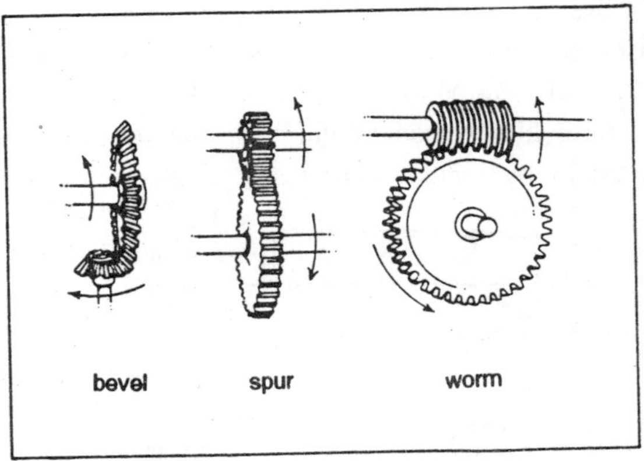

Fig. G-2. Types of gears.

Gear. Toothed wheel, cylinder, or cone, usually mounted on a shaft, that transmits motion from one part of a machine to another. When the teeth of two gears are meshed, the turning of one shaft will cause the other shaft to rotate. By meshing two gears of different diameters, a variation in both speed and torque is obtained.

Geiger-Nuttal law. The principle that the range *(R)* of a particle emitted in alpha decay is related to the decay constant(λ) by the equation log $\lambda = C$ or $R + B$, where C and B are constants.

Geissler tube. A glass or quartz tube in which a discharge is struck between two electrodes in a low pressure of gas. The electrodes are contained in bulbs connected by a length of capillary tube in which the discharge is intensified. The device is used as a light source for sectroscopy. [After Henry Geissler (1814-79), German instrument maker.]

Gel. A colloidal suspension in which the dispersed particles link together, forming a jelly-like mass. Gels can be produced by coagulation of lyophilic sols. They are often thixotropic.

Gelatin. A colourless or yellow solid obtained in sheets. flakes, or granules. It is a protein, made by boiling skin, tendons, or bones with water, and is used in jellies, films, and adhesives.

Gelatin or **animal jelly.** Foodstuff obtained from connective tissue (found in hoofs, bones, tendons, ligaments, and cartilage) of vertebrate animals by the action of boiling water or dilute acid. It is largely composed of the protein collagen. Pure gelatin is brittle, transparent, colorless, tasteless, and odorless. It dissolves in hot water and congeals when cooled Gelatin is widely used to give food a proper consistency, in photographic emulsions, and as a coating for pills. Vegetable gelatin, or Agar, is made from seaweed.

Gelignite. A high explosive consisting of a mixture of nitroglycerine with nitrocellulose, potassium nitrate, and wood pulp.

Gell-Mann, Murray, 1929-. American theoretical physicist; b. N.Y.C. In 1953 he and, independently, the Japanese team of

T. Nakano and Kazuhiko Nishijima proposed the concept of "strangeness" to account for certain particle-decay patterns. In 1961 Gell-Mann and Yuval Ne'eman independently introduced the "eightfold way," or SU(3) symmetry, a table like ordering of all subatomic particles. The 1964 discovery of the omega-minus particle, which filled a gap in this ordering, brought the theory wide acceptance, and led to Gell-Mann's being awarded the 1969 Nobel Prize for physics. In 1963 Gell-Mann and George Zweig independently postulated the existence of the Quark, an even more fundamental particle with a fractional electric charge.

Gem. Commonly, a mineral or organic substance that is cut and polished and used as an ornament. The qualities sought in gems are beauty, rarity, and durability. The unit of weight used for gems is the metric carat (200 mg). Gems are usually cut to bring out their colour and brilliance and to remove flaws, and gem cutting is a distinct trade. The precious stones are Diamond, some forms of Corundum (e.g., Ruby and Sapphire), and Emerald. The chief semiprecious stones include Amethyst, Aquamarine, Garnet, Jade, Moonstone, Opal, Quartz, Topaz, Tourmaline, and Turquoise. The organic gems are Amber, Coral, Pearl, and jet. Synthetic gems produced using the chemical elements of natural stones include diamonds, emeralds, rubies, and sapphires. See also Agate; Beryl; Bloodstone; Jasper; Lapis Lazuli, Serpentine.

Gene. Ultimate unit by which inheritable characteristics are transmitted to succeeding generations in all living organisms. Genes are contained by, and arranged along the length of, the Chromosome. The gene is composed of deoxyribonucleic acid (see DNA.) Each chromosome of each species has a definite number and arrangement of genes, which govern both the structure and metabolic functions of the cells and thus of the entire organism. They provide information for the synthesis of enzymes and other proteins and specify when these substances are to be made. Alteration of either gene number or arrangement can result in Mutation, a change in the inheritable traits. *See* Genetics. *See* Genetic Engineering.

General relativity. The general theory of relativity.

Generator. 1. (dynamo). A machine for producing an electric current by electromagnetic induction. Generators convert mechanical energy into electrical energy and depend on relative motion of a conductor and magnetic field, so that a current is induced. In a simple alternating-current device the coil is rotated in a constant magnetic field produced by a *field magnet.* This coil, the *armature,* is connected to conducting cylindrical *slip rings* on the driving shaft and the current is taken off by carbon brushes. The current changes direction, the frequency depending on the rate of rotation. In small a.c. generators the armature is a steel cylinder wound with three coils at 120°. The field is produced by an electromagnet, supplied by a direct current, with curved pole pieces. Such generators produce three a.c. output differing in phase by 120° .

Direct current generators have slip rings made in segments, forming a *commutator.* As the coil rotates the current is taken from different parts of the commutator so that it always flows in the same direction. A single coil gives a half-wave output. If the armature contains a number of coils. The resulting output can be direct current, with a small ripple. Another type of d.c. generator is the Faraday disc.

All generators consist of two parts: the *stator,* which remains stationary, and the *rotor,* which moves. Usually, the stator produces the field and the current is induced in the rotor. In very large a.c. generators this is reversed: the magnetic field is produced by rotating field coils and the current is induced in a stationary armature. This arrangement has the advantage that brushes are used to supply the relatively small direct currents to the field coils, and are not required for the large generated currents.

2. Any of various devices for producing electrostatic charge, as in the Van de Graaff generator.

3. Any of various pieces of chemical apparatus for producing gases for laboratory use, as in Kipps apparatus.

4. A line on the surface of a cone or cylinder lying in the same plane as the axis, such that the surface could be generated by rotating the line about the axis.

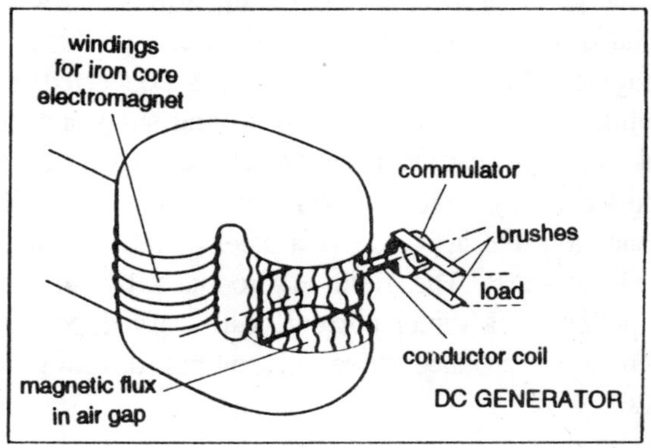

Fig. G-3. *Generator:* The rotation of the conducting coil of wire in a magnetic field causes a current to be induced in the coil. In the DC generator, a split-ring commutator changes the direction of the current with each half rotation, resulting in a pulsing direct current.

Generator. Electrical device used to convert mechanical energy to electrical energy. It operates on the principle of electromagnetic Induction. The generator moves a conductor through a magnetic field and directs the current produced by the induced voltage to an external circuit. In the simplest generator, the conductor is an open coil of wire rotating between the poles of a permanent magnet. During a single rotation, one side of the coil passes through the magnetic field first in one direction and then in the other, so that the induced current is alternating current (AC), moving first in one direction, then in the other. Each end of the coil is attached to a separate metal slip ring that rotates with the coil. Brushes resting on the slip ring pass the current to the external circuit. To obtain direct current (DC) i.e., current that flows in only one direction, a commutator is used in place of slip rings. The commutator is a single slip ring split into left and right halves that are insulated from each other and attracted to opposite ends of the coil. Current leaves the generator through the brushes in only one direction and pulsates from no flow to maximum flow and back again. In practice, generators have many coils and several magnets. The whole assembly carrying the coils is called the armature, or rotor; the stationary parts constitute the stator. Except for magnetos, which use permanent magnets, AC and DC generators use electromagnets. AC generators are often called alternators.

Genetic engineering. Group of new research techniques that manipulate the DNA (genetic material) of cells. The genesplicing technique, which produces recombinant DNA, is a method of transporting selected genes from one species to another. In this technique, the genes, which are actually portions of molecules of DNA, are removed from the donor (insect, plant, mammal, or other organism) and spliced into the genetic material of a virus; then the virus is allowed to infect recipient bacteria. In this way the bacteria become recipients of both viral and foreign genetic material. When the virus replicates within the bacteria, large quantities of the foreign as well as viral material are made. The introduction into bacteria of foreign genetic material is an important tool for the study of gene structure and regulation, but it also presents risks that some genetic material would no longer be under natural control. The recipient bacterium in use is a modified form of *E. coli,* a natural inhabitant of human intestines. Other techniques include cell fusion which has made the mapping of human genes possible, and nuclear transplantation, valuable to the study of factors controlling embryological development. The enzymes used in gene splicing were first studied by microbiologists Werner Arber, of the Univ. of Basel, Switzerland, and Daniel Nathan and Hamilton

O. Smith, both of Johns Hopkins Univ., Baltimore, who shared the 1978 Nobel Prize in medicine for their work.

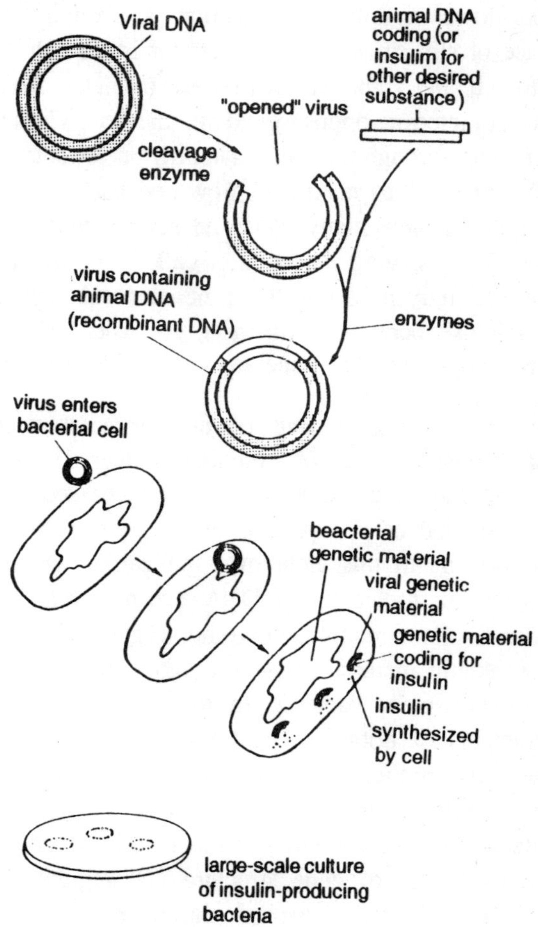

Fig. G-4. Genetic engineering: Gene-splicing techniques.

Genetics. Scientific study of heredity. The science arose in 1900, with rediscovery of Gregor Mendel's work on traits that are inherited as if each were a separate, independent unit. Geneticists probing the physical basis of the transmission of inherited characteristics have studied the unit of inheritance, the Gene;

the small chemical differences in genes that are expressed as different versions of the same trait (alleles); the array of genes along the Chromosome in the cell nucleus; and the set of genes characteristic of each species (the genome). Since the discovery (1953) of the structure of DNA, work on Nucleic Acids has explained how genes determine all life processes by directing the synthesis of all cell proteins. It has also explained Mutation as alterations in gene or chromosome structure. See also Genetic Engineering.

Genocide. Systematic destruction by a government of a racial, religious, or ethnic group. The term was coined in 1944, with particular reference to the attempt of Nazi Germany in the 1930s and 40s to destroy European Jewry (see Holocaust; National Socialism). Other instances of genocide exist, however, such as the massacre of Armenians in Ottoman Turkey (1894-96). A 1949UN convention defines the crime of genocide and provides for prosecution by national or international courts. The U.S. has never ratified the convention, claiming that the provision for international tribunals and the principle of personal liability for national actions violate the concept of Sovereignty.

Geocentric. 1. Having the earth at the centre.

2. Measured with reference to the earth.

Geochemistry. The study of the chemical composition and reactions of the substances making up the earth.

Geochemistry. Study of the chemistry of the earth, especially the study of the absolute and relative abundances of the elements and their distribution and movement. Such studies have provided insights into the evolution of the oceans and atmosphere, the ages of rocks, the chemical evolution of life, and, more recently, the effects of the massive introduction of pollutants into the environment.

Geodesic. A line on a surface that is the shortest distance between two points on the surface. On a sphere, for example, the geodesics are great circles.

Geodesy or **geodetic surveying.** Subdivision of Geophysics that determines the earth's size and shape, the position of points on the earth's surface, and the dimensions of areas so large that the curvature of the earth must be taken into account. In a process called triangulation, a base line between two points, many miles apart, is measured, and a third point is determined by the angle it makes with each end of the base line. In recent years, triangulation has been done increasingly with geodetic satellites in space.

Geologic eras. Major units of geologic time. For the purpose of dating rock formations and the fossils contained within them, the earth's history has been divided into eras, which are further subdivided into periods and, in some cases, epochs.

Geology. Science of the Earth's history, composition, and structure. Branches include mineralogy. Petrology, geomorphology (the origin of landforms and their modification by dynamic processes). Geochemistry, Geophysics, sedimentation, structural geology, economic geology, and engineering geology. Historical geology includes stratigraphy (the interrelationships of layered Rocks) and paleontology (the study of Fossils). Geological observations have been made since ancient times, but modern geology began in the 18th cent., notably with James Hutton's doctrine of Uniformitarianism, which, in its opposition to Catastrophism, laid the groundwork for much of modern geological science. The systematic survey in the mid-20th cent. of the Ocean floors brought radical changes in concepts of crustal evolution (see Plate Tectonics).

Geomagnetism. *See* terrestrial magnetism.

Geometrical isomerism. A type of stereoisomerism in which two isomers differ in the relative positions of two differ in the

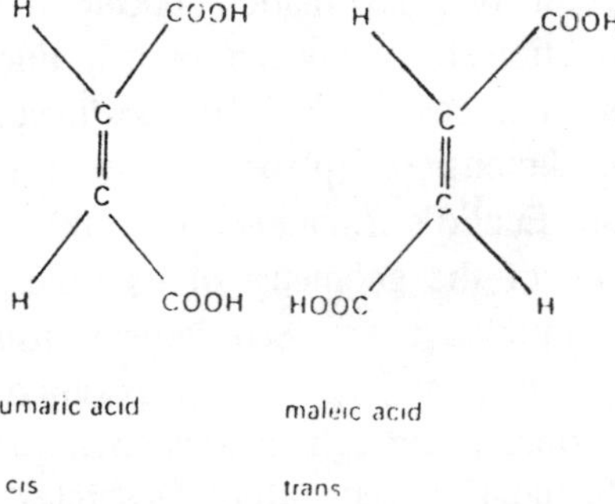

Fig. G-5. Geometrical isomers.

relative positions of two groups about a double bond. Fumaric and maleic acids are examples of geometrical isomers. In fumaric acid both carboxylate groups are on the same side of the double bond (cis) while maleic acid has the groups on opposite sides *(trans)*. This type of behaviour is often called *cis-trans isomerism.* It can also be found in certain ring compounds. Square-planar and octahedral inorganic compexes also have geometrical isomers depending on whether two ligands are adjacent or opposite.

Geometrical optics. *See* optics.

Geometric mean. The *n*th root of the product of *n* numbers. Thus, the geometric mean of *a,b,c,* and *d* is $(abcd)^{1/4}$

Geometric progression. A sequence in which each term is the product of the preceding term and a fixed number. A geometric progression has the form $a, ar, ar^2, ... ar^{n-1}$, where r is the *common ratio* and n is the number of terms. The *geometric series* formed from this sequence has a sum equal to $[a(1 - r^n)]/(1 - r)$. If $-1 < r < 1$ the series converges to $a/(1 - r)$. *Compare* arithmetic progression.

Geometry. The mathematical study of shape. *Plane geometry* is the study of figures and curves in two dimensions: *solid geometry* is their study in three dimensions. *See also* Euclidean geometry, non-Euclidean geometry, analytic geometry.

Geometry. Branch of Mathematics concerned with the properties of and relationships between points, lines, planes, figures, solids, and surfaces. Elementary Euclidean Geometry of two and three dimensions (plane and solid geometry) is based largely on Euclid's *Elements (c.* 300 B.C), a systematic presentation of the geometry of its time. Although Euclid's basis assumptions have been judged insufficiently rigorous by David Hilbert and others, his axiomatic method of proof has been adopted throughout mathematics and in other fields as well. In the 17th cent., Rene Descrates invented Cartesian Coordinates to express geometric relations in algebraic form; he thus founded Analytic Geometry, from which developed algebraic geometry, concerned with geometrical objects defined by algebraic relations between their coordinates. Other branches of geometry include discriptive geometry, which is concerned with the two-diferential geometry, in which the concepts of the Calculus are applied to curves, surfaces, and other geometrical entities; projective geometry, which is concerned with those properties of geometrical figures that remain unchanged under projection, e.g., from one plane to another; and Noneluclidean Geometry. Topology, perhaps the most general type of geometry, is often considered a separate branch of mathematics.

Geophysics. The study of the physical properties and behaviour of the earth, including such topics as seismology, oceanography, atmospheric electricity, gravitational properties, and terrestrial magnetism.

Geophysics. Study of the structure, composition, and dynamic changes of the Earth and its Atmosphere, based on the principles of physics. Applied geophysics use seismic, electrical, gravimetric, magnetic, and radiometric techniques for geological exploration and prospecting.

Geranium. Herb or shrub of the family Geraniaceae. More specifically, the name refers to the popular garden and greenhouse geranium (geneus *Pelargonium),* grown for its colourful flowers and ornamental, sometimes scented foliage. Its long. Beak-shaped fruits give it the name stock's bill. The family also includes the true geraniums (genus *Geranium)* and the heron's-bills (genus *Erodium).* Geraniums are also cultivated for the aromatic oils extracted from their foliage and flowers.

Geriatrics. Study of the medical problems of the aged. Aging cells are more susceptible to the accumulation of Calcium. Cholesterol, and other substances that may cause tissue deterioration, and thus the elderly are more prone to disease and decreased physiological functioning. Many disabilities of aging are related to Circulatory System deterioration, but studies have shown that proper nutrition and exercise contribute to prolonging good health and circulation. The general study of old age is known as Gerontology.

Germanium. Symbol: Ge. A brittle crystalline grey-white metalloid element found in some zinc ores and some coals. It is extracted by producing the volatile tetrachloride, hydrolysing it to give the oxide, and then reducing this with hydrogen. The principal use is in semiconductors, for which the pure material is obtained by zone refining. The chemistry of germanium is similar to that of silicon, although germanium is rather more reactive. In most of its compounds its valency is 4 although some divalent salts are known, such as $GeCl_2$ and GeS. A.N. 32; A.W. 72.59; m.pt. 937.4°C; b.pt. 2830°C; r.d. 5.32.

Germanium. (Ge), Semimetallic element, isolated from argyrodite by Clemens Winkler in 1886. Gray-white, lustrous, and brittle, it is chemically and physically similar to Silicon. It is used as a Semiconductor in Transistors and integrated circuits. The oxide, transparent to infrared radiation, is used in optical instruments. See Element (table), Periodic Table.

Germination. Process by which the plant embryo within the Seed resumes growth after a period of dormancy and the seedling emerges. Food stored in the endosperm or in cotyledons provides energy in the early stages of growth until the seedling can make its own.

Gerontology. Study of old age and the aging process, concerned with the physical aspects of aging as well as with the special economic and social problems of the elderly. Gerontology emerged as a major field of study in the 20th cent. as the steady increase in average life expectancy swelled the numbers of older people in the population, especially in developed countries. The medical specialty that focuses on the health and diseases of the elderly is called Geriatrics.

Gesneria. Herb or shrub, chiefly tropical and subtropical, of the family Gesneriaceae, cultivated for its showy, often tubular blossoms. The family includes the African Violets; the African Cape primrose (genus *Streptocarpus);* and the gloxinia *(Sinningia speciosa)* of Brazil, not to be confused with the genus *Gloxinia,* which is not cultivated.

Gestation. In mammals, the period between conception and birth in which the developing young animal, or embryo, is carried within the uterus. In humans, gestation is usually called Pregnancy.

Getter. A substance used for removing gas from a vacuum tube by adsorption or chemical reaction. Most getters are pure metals. Magnesium, for example, is used in thermionic valves, being heated to combine with residual traces of oxygen and nitrogen after the tube has been sealed. In experimental vacuum equipment at low pressures the getter can be continuously deposited as a clean evaporated film, which adsorbs the gas. A similar process occurs in the getter-ion pump (*see* ion pump).

Getter-ion pump. *See* ion pump.

GeV. Gigaelectronvolt; 10^9 electron-volts. In the U.S. *BeV* is used *B* stands for an American billion.

Geyser. Hot Spring from which water and steam are ejected periodically to heights ranging from a few to several hundred feet. The generally accepted explanation is that rain-water, collected under pressure in hot rocks below ground, turns partly to steam and causes the water above it to overflow, thereby reducing the pressure and forcing an eruption. Notable geysers are found in Iceland; North Island, New Zealand, and Yellowstone National Park, U.S. (see National Parks, table).

Giant. A giant star; a star of enormous diameter, low mean density, and a high absolute magnitude.

Giant planet. Any of the group of outer planets, Jupiter, Saturn, Uranus, and Neptune, distinguished from the other terrestrial planets by their large size and mass and by a relatively low density.

Gibbous. Denoting the moon when its phase is between a half moon and a full moon.

Gibbs, Josiah Willard. 1839-1903, American mathematical physicist; b. New Haven, Conn. He was professor of mathematical physics at Yale. Gibbs's studies in physical chemistry and thermodynamics have had profound effect on industry, notably in the production of ammonia. He formulated the concept of chemical potential, was influential in developing vector analysis, and did important work in statistical mechanics.

Gibbs free energy (Gibbs function). Symbol: G. A free-energy function of a thermodynamic system equal to the enthalpy minus the product of the entropy and the thermodynamic temperature: i.e. $G = H - TS$. At constant temperature and pressure the value of G is a minimum when the system is in equilibrium. These are the conditions most usually used in chemical reactions.

Gilbert. Symbol: Gb. A CGS-electromagnetic unit of magnetomotive force equal to the magnetomotive force produced by a current of 40π amperes (4π abamperes) passing through a single turn of a coil. It is equivalent to 10/4π ampere-turns.

Gilbert, William, 1537-1603. English scientist and physician. Noted for his studies of Electricity and Magnetism, he coined the word *electricity* and was the first to distinguish between electric and magnetic phenomena. In his *De Magnete* (100) he described methods for strengthening natural magnets (lodestones) and for using them to magnetize steel rods by stroking, and he concluded that the each act like a giant magnet with its poles near the geographic poles. From 1600 he was president of the College of Physicians and court physician to Elizabeth I and James I.

Gill. External respiratory organ of most aquatic animals, the basic function of which oxygen-carbon-dioxide exchange. Gill structure and location vary among animals of different groups. In fishes, gills are located at the rear of the mouth and contain capillaries; in higher aquatic invertebrates, they protrude from the body surface and contain extensions of the vascular system; in mollusks, they are inside the mantle cavity; in aquatic insects, they occur as projections from the walls of the air tubes. In amphibians, gills are usually present only in the larval stage; in higher vertebrates, they occur merely as rudimentary, nonfunctional gill slits, which disappear during embryonic development.

Gin. Spirituous liquor distilled mainly from fermented cereals and flavoured with juniper berries. Of Dutch origin, it is now chiefly made in England and the U.S. Types of gin include London (highly distilled), Old Tom (sweetened), and sloe (flavored with sloes instead of juniper berries).

Ginger. Common name for perennial herbs of the tropical and subtropical family Zingiberaceae. Many are important for their aromatic oils. Ginger (*Zingiber officinale*) is cultivated for its root, which is candied or dried for medicines and spice. Turmeric (*Curcuma longa*) and the seeds of cardamom (*Elettaria cardamomum*) are similarly used, and often combined with ginger to make a curry.

Gingivitis. Inflammation of the gums, characterized by red, swollen, spongy gums that bleed easily. It may be acute, chronic, or recurrent. Chronic gingivitis usually results from bacteria, although other factors, such as poor dentition, prolonged use of the drug phenytoin, vitamin C deficiency, or Diabetes, may also contribute to it. It left untreated, gingivitis can lead to · Periodontitis.

Ginkgo or **Maidenhair tree.** Deciduous tree (*Ginkgo biloba*) with fan-shaped leaves. A sacred plant in its native China and Japan, ginkgo is a "living fossil," the only remaining species of a large order (Ginkgoales) of Gymnosperms that existed in the Triassic period. Today the ginkgo is valued in the U.S. and Europe as an avenue tree because of its exceptional tolerance for smoke, low temperature, and minimal water supply (although the female plant, with its malodorous fruit, is not as desirable an ornamental as the male). Ginkgo seeds are esteemed as a food in the Orient.

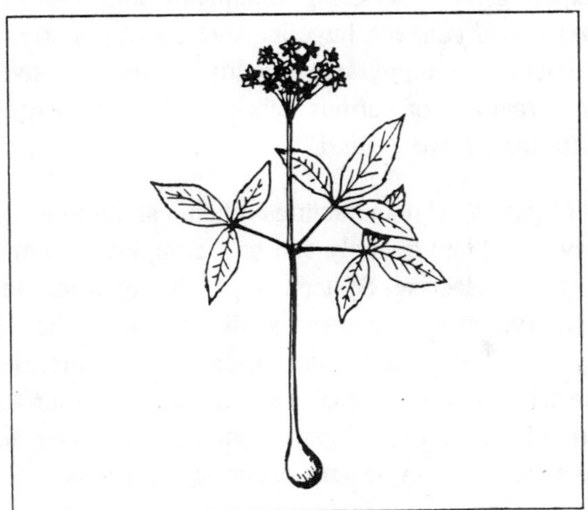

Fig. G-6. *Dwarf ginserg,* Panax trifolium.

Ginserg. Plant (genus *Panax*) of the family Araliaceae. True ginserg (*P. schinseng*) is prized by the Chinese for its curative properties. It and a North American ginseng (*P. quinquefolius*), used as a substitute for true ginseng, have been nearly exterminated by commercial exploitation.

Giraffe. African ruminant Mammal (*Giraffa camelopardalis*) living in open Savanna S of the Sahara. The tallest animal (upto 18 ft/5.5 m), the giraffe browses in the tops of Acacia and Mimosa trees. The legs and neck are elongated; the skin is patterned in large, sandy-to-chestnut spots on a lighter background. Giraffes travel in small herds.

Glacial acetic acid. Pure acetic acid, free from water.

Glacier. Mass of ice formed in high mountains and polar regions by the compacting of snow and kept in constant movement by the pressure of the accumulated mass. The four main types are *valley, or mountain, glaciers,* tongues of ice moving from mountain snowfields into stream valleys; *piedmont glaciers,* formed by the spread of one or the convergence of several valley glaciers; *ice caps,* flattened, somewhat dome-shaped galciers covering mountains and valleys alike; and *continental glaciers,* huge ice sheets that give rise to Icebergs. Glaciers alter topography greatly by their erosive action, by their transport of various debris, and by the various forms of Drift they leave behind.

Gland. Organ that manufactures chemical substances. A gland may vary from a singly cell to a complex system. The glands of the Endocrine System, e.g., the thyroid, adrenals, and pituitary, secrete hormones directly into the bloodstream. Sweat, salivary, and other so-called exocrine glands secrete substances into external or internal body surfaces, usually through ducts. Mixed glands such as the Liver and Pancreas have both endocrine and exocrine functions.

Glanders. Highly contagious, fatal bacterial disease of horses, mules, and donkeys that can be transmitted to humans. The bacterium, *Actinobacillus mallei,* primarily infects the skin, lungs, and nasal membranes, causing lumps, or nodules, to form in the infected area. These nodules enlarge, from ulcers, and release bacteria-laden pus to other parts of the body. Infected animals must be destroyed to prevent the spread of the disease.

Glass. Hard substance, usually brittle and transparent, composed chiefly of silicates and an alkali fused at high temperatures. Metallic oxides impart color. In prehistoric times objects were fashioned form natural glass such as obsidian (avolcanic substance) and rock crystal (a transparent quartz). The oldest extant manufactured glass is form Egypt, c.2000 B.C. Many types were made in Roman times, but little is known of European glassmaking from the fall of Rome until the 10th cent., when Stained Glass appeared. Methods have changed little since ancient times. The materials are fused at high temperatures in seasoned fireclay containers, boiled down, skimmed, and cooled several degrees; then the molten glass is ladled or poured into molds and pressed, or it is blown or drawn. The shaped glass is annealed to relieve stresses caused by manipulation, then slowly cooled. Until the 17th cent. The finest glass was made in Venice; later France and England became centers of glassmaking. In the 20th cent. many new types have been developed, including fiberglass and safety glass, but some uses of glass have been superseded by Plastic. Despite mass production, glassmaking by hand remains a valued art.

Glass. 1. A hard brittle transparent or translucent substance made by fusing together a mixture of sand (SiO_2), sodium carbonate, and lime. This gives a mixture of calcium and sodium silicates called *soda glass*, which is the common form of glass used in bottles, windows, etc. Other specialized glasses can be made in which part of the sodium or calcium is replaced by metals such as barium and lead. Lead glass, for example, has a high refractivity. Similarly some of the silicon atoms can be replaced by boron atoms forming *borosilicate glass*, which has a lower thermal expansion (*see* Pyrex). Coloured glass is made by adding coloured metal oxides to the fused mixture.

Glass is an amorphous substance: i.e. the atoms in the solid do not have the regular pattern characteristic of a crystalline solid. Instead they have a fairly random arrangement similar to that in a liquid. Thus glass does not show a sharp melting point but softens over a temperature range. It is best described as a supercooled liquid.

2. Any amorphous solid produced by cooling a liquid.

Glauber's salt. *See* sodium sulphate.

Glaucoma. Disease of the Eye characterized by an excess of fluid within the eyeball, causing increased pressure on the retina and impairment of vision ranging from slight abnormalities to Blindness. It most commonly develops gradually after the age of 40 but can sometimes occur abruptly. It is treated with drugs (such as pilocarpine) that decrease intraocular pressure.

GLC. *See* chromatography.

Glider. Type of aircraft resembling an airplane but having at most a small auxiliary propulsion plant and often no means of propulsion at all. Modern gliders have very slender wings and streamlined bodies. The unpowered glider is usually launched by an elastic shock cord, a rope, or a cable attached to its front and pulled by a launching crew, a tow car, or a tow plane; it uses gravity keep flying and updrafts of air to gain altitude. The type of craft built especially for soaring and sustained flight is called a sailplane.

Glow discharge. A type of electrical discharged occurring between two electrodes in a low pressure of gas. The discharge can be maintained in pressures of a few mmHg and consists of a number of dark and bright bands in the region between the electrodes. Most of the potential drop across the discharge occurs between the cathode and a small luminous region close to the cathode called the *cathode glow*. This region is separated from the cathode by a narrow band, the *Crookes dark space*. Another luminous region, the *positive column*, extends from the anode: it is separated from the cathode glow by another dark region, the *Faraday dark space*. At lower pressures the positive column forms into a set of alternate light and dark bands, called *striations*.

The characteristic features of the glow discharge are the nonuniform fall in voltage with distance down the tube, the relatively high current (a few milliamps), and the fact that the voltage drop across the discharge is independent of the current. This last feature makes the glow discharge suitable for use as a voltage stabilizer. *See also* Townsend discharge.

Glucagon. Polypeptide Hormone secreted by the islets of Langerhans in the Pancreas. It tends to counteract the action of Insulin, i.e., it raises the concentration of glucose in the blood, essentially by promoting Glycolysis.

Glucose. A colourless crystalline sugar, $C_6H_{12}O_6$, present in many plants and in sucrose and starch. Carbohydrates and other sugars are broken down to glucose in the body. The natural form is optically active and is also called D-glucose or *dextrose*. M.pt. 146°C; r.d. 1.54.

Glucose, Dextrose, or Grape Sugar (Empirical formula: $C_6H_{12}O_6$, white crystalline Sugar, somewhat less sweettasting than Sucrose (table sugar), it is found in fruits and honey. Glucose is the major source of energy in animal metabolism. It requires no digestion prior to absorption into the bloodstream. A monosaccharide, glucose can be obtained by Hydrolysis of a variety of more complex carbohydrates, e.g., Maltose, Cellulose, or Glycogen. It is commercially made from cornstarch and is used in sweetening candy, chewing gum, jellies, and various foods. Glucose present in urine may be a symptom of Diabetes.

Glyceride. An ester formed between glycerol and a fatty acid. Glycerol, $CH_2OHCH(OH)CH_2OH$, is a trihydric alcohol and one molecule can react with either one, two, or three molecules of acid. Natural fats and oils are *triglycerides*—esters in which all three hydroxyl groups have been substituted (see illustration). The three fatty acid molecules may be identical (*simple glyceride*) or different (*mixed glyceride*). Many different fatty acids are found in natural fats and oils: they are generally classified into saturated and unsatruated acids. Saturated fatty acids, such as stearic and palmitic acids, have a straight-chain alkyl group attached to a carboxyl group. In unsaturated acids, such as oleic and linoleic acids, the hydrocarbon chain contains one or more double bonds. Fats are solid at room temperature and contain a higher proportion of saturated acids: oils are liquid and contain more unsaturated acids.

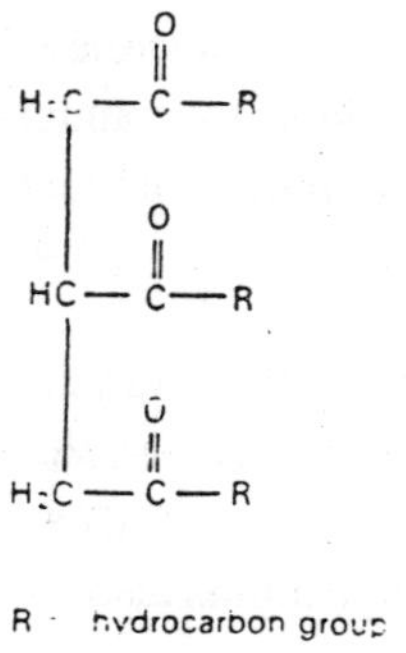

R · hydrocarbon group

Fig. G-7. Triglyceride.

Glycerol (glycerine). A colourless syrupy liquid, $CH_2OHCHOHCH_2OH$, with a sweet taste, obtained as a by-product in the manufacture of soaps. It is used as a solvent, humectant, and plasticizer. M.pt. 18°C;b.pt.290°C;r.d.1.3.

Glycine (aminoethanoic acid). A white sweet-tasting crystalline solid, $NH_2CH_2{-}COOH$, made by alkaline hydrolysis of gelatin. The principal amino acid in sugar cane, glycine is used in organic synthesis, medicine, and as a nutrient. M.pt.233°C (decomposes); r.d. 1.2.

Glycogen. Highly branched Polymer of Glucose that is made and stored in the Liver and Muscle cells of humans and the higher animals and in the cells of lower animals. During short periods of strenuous activity, energy is released in the muscles by direct conversion of glycogen to lactic acid. See also Carbohydrate.

Glycolysis. Process in all higher animals and most microorganisms for the degradation of glucose. Beginning with a single molecule of glucose, glycolysis is a series of chemical reactions requiring eleven different Enzymes and eventually yielding two molecules of lastic acid, which then enter the Citric Acid Cycle. The reactions of glycolysis also generate the high-energy substance Adenosine Triphosphate (ATP).

Glyptal resin. *See* polyester resin.

Gnat. Small, two-winged Insect (suborder Nematocera) of the order Diptera (see Fly). Gnats include crane flies, midges, black flies, Mosquitoes, and others. Gnats often assemble in large matings warms. Some gnats are suspected carriers of pinkeye and yaws.

Gneiss. Coarse-grained, imperfectly layered metamorphic Rock, characterized by alternating dark and light bands that differ in mineral content. Gneisses result from the metamorphism of many igneous or sedimentary rocks and are the most common types of rocks found in Precambrian regions.

Gnu or **Wildebeest.** Large Antelope (genus *Connochaetes*) living in herds on African grasslands. A swift runner, it has hindquarters like those of a horse and a head and humped shoulders like those of a buffalo. A beard, mane, long tail,and large curving horns are characteristic of both sexes. The brindled gnu, or blue wildebeest (*C. taurinus*), weighs c.500 lb (225 kg) and stands $4^1/_2$ft (135 cm) at the shoulder.

Goat. Ruminant Mammal (genus *Capra*) of the Cattle family with hollow horns, coarse hair, and a characteristic "bread," closely related to the Sheep. Found in mountainous terrain and arid climates worldwide, goats live in herds and feed on grass and shrubs. They have been bred for centuries for their highly nutritious milk and valuable hair and wool, the source of mohair and cashmere. Wild goats include the Ibex.

Goatsucker. Medium-sized primarily nocturnal Bird (order Caprimulgiformes) found in temperate and tropical regions, including the whippoorwill. Goatsuckers have long pointed wings, weak feet, and small gaping bills fringed with bristles. Most are brown, gray, or black and have monotonous songs. They feed mainly on insects. All fly at night except the nighthawk, which is active at dawn and dusk. The whippoorwill, common in the E U.S., hibernates during the winter instead of migrating.

Godel, Kurt (go dal), 1906-78. Czech- American mathematician and logician. He came to the U.S. in 1940. Godel is best known for his work in mathematical logic, particularly his proof (1931) of a theorem stating that the various branches of mathematics are based in part on propositions that are not provable within mathematics itself, although they may be proved by means of logical systems external to mathematics.

Gold. Symbol: Au. A soft very malleable and ductile heavy yellow transition element found native and as the telluride. The metal is obtained by sluicing, by amalgamation with mercury, or by dissolving in potassium cyanide followed by reduction of the resulting cyanoaurate. A certain amount is recovered from the anode sludge produced in electrolytic refining of copper. Gold is used in coinage, jewellery, and, in the form of gold leaf, in decorative gilding. Chemically, the metal is less reactive than copper or silver, being unaffected by oxygen, sulphur, and most acids. The element is attacked by the halogens and by aqua regia. Gold does not have typical ionic salts; most of its compounds are covalently bonded and many complexes exits. It forms both gold I (*aurous*) and gold II (*auric*) compounds. A.N. 79; A.W. 196-97; m.pt. 1063°C; b.pt. 2966°C; r.d. 9.32; valency 1 or 3.

Gold (Au). Metallic element, known since prehistoric times. Chemically inactive, this very ductile and most malleable Metal can be beaten into thin sheets of gold leaf. Only silver and copper conduct electricity better. Gold is usually hardened by alloying with other metals; it also often occurs in nature as an alloy. The gold content of an alloy is stated in carats; by definition, pure gold is 24 carats, and thus 75% gold alloy is 18 carats. Gold is found widely distributed, mostly in metallic form as duct, grains, flakes, or nuggets. It occurs in quartz veins or lodes, usually in association with silver or other metals, and in a alluvial placer deposits. Possibly the first metal used by humans, gold was valued for ornaments, and magical powers were attributed to it. Alchemists of the Middle Ages tried to transmute baser metals into gold (see Alchemy). The search for gold stimulated European exploration of the Western Hemisphere. For discussion of its monetary function, see Bimetallism; Coin; Money. *See also* Element (table); Periodic Table.

Gold chloride (auric chloride, gold trichloride). The only stable chloride of gold, $AuCl_3$. It is a red crystalline solid, prepared by the action of aqua regia on gold. A hydrated form. $HAuCl_4 4H_2O$, is *chloroauric acid,* which is used in photography, gold plating, and ceramics.

Golden section. The division of a line or area so that the ratio of the larger to the smaller part is equal to the ratio of the whole to the larger. If a point O divides a line AB such that AO/OB = AB/AO, the ratio AO/OB is the *golden mean.* A rectangle having sides in this ratio is a *golden ractangle,* which can be divided into a square and another golden ractangle. Proportions based on golden sections appear in many paintings; they are considered to be particularly aesthetically pleasing.

Golden calf. In the Bible, an idol erected by the Israelites on several occasions. Aaron made one while Moses was on Mt. Sinai. Ex. 32. Jeroboam I made two, and Hosea denounced a calf in Samaria. I Kings 12-; Hosea 8. A bull cult was widespread in Canaan at the time of the Israelite invasion.

Golden Fleece. In Greek mythology, magic fleece of the ram that carried Phrixus and Helle from Boeotia. Helle fell into the sea (which became the Hellespont), but Phrixus arrived safely in Colchis. He then sacrificed the ram, which became the constellation Aries. Its fleece, hung in a wood guarded by a dragon, was later sought by Jason and the Argonauts.

Golden retriever. Large Sporting Dog; shoulder height, c. 23 in. (58 cm); weight, 60-75 lb (27.2-34.1 kg). Its water-repellent double coat is golden-brown. Developed in Scotland in the 19th cent. mainly to hunt waterfowl, this hardy breed it also used as a Guide Dog.

Goldish. Freshwater Fish (genus *Carassius)* popular in home aquariums. Native to China, it was domesticated centuries ago from its wild form, an olive-coloured carplike fish up to 16 in. (40 cm) long. Marketed goldfish are 1 to 4 in. (2.5 to 10 cm) long and most commonly orange in colour; breeders have developed bizarre varieties with unusual tails, double or triple fins, bulging eyes, and a range of colors.

Gold point. *See* temperature scale.

Goldschmidt process. A process for extracting metals from their oxides by reduction with aluminium. The reaction occurring is the same as that in thermite.

Goniometer. An instrument for measuring the angles between the faces of crystals.

Goose. Large wild or domesticated swimming Bird related to the Duck and Swan. Strictly speaking, the term goose applies to the female and *gander* to the male. In North America the wild, or Canada, goose *(Branta canadensis)* is known by its honking call and V-shaped, migrating flocks in spring and fall. Geese have been domesticated since ancient times; the popular Toulouse, or gray, goose is descended from the European graylag *(Anser anser).*

Gopher or **pocket gopher.** Burrowing Rodent (family Geomyidae) found in North and Central America. The gopher is gray, buff, or dark brown; its combined head and body length is 5 to 12 in. (13 to 30 cm). It has extremely long teeth and large claws. The name *pocket gopher refers* to the furlined pouches that open on the outsides of its cheeks, used to carry food and nesting material.

Gorilla. Largest Ape *(Gorilla gorilla),* native to the forests of equatorial W. Africa. Males range from 5 to 6 ft (150 to 190 cm) in height and weigh c. 450lb. (200 kg) in the wild; they have prominent sagittal crests and brow ridges and enormous canine teeth. Females are about half the size of males and their features are less developed. Their shaggy coats are brown or black. Though extremely muscular and powerful, gorillas are quiet and retiring; they are chiefly vegetarian.

Gourd. Common name for some members of the Cucurbitaceae family, of tropical, subtropical, and temperate regions. The family is known for its many edible and otherwise useful plants, almost all annual herbs that grow as vines. Among these are *Cucurbita* species, including the Pumpkin and summer squashes (varieties of *C. pepo)* and the winter squashes (*C. maxima);* and *Cucumis* species, including the Cucumbers, Gherkins *(C. anguria),* and all Melons except the Watermelon. The edible fruit of the loofah, or California okra *(Luffa cylindrica),* is dried for the inner fibrous network, which is used as a scrubbing sponge. The term *gourd* is applied to those members of the family whose fruits have hard, durable shells used for ornament and as utensils, e.g., cups, dippers, and bowls. The Old World genus *Lagenaria* includes the calabash, dipper, and bottle gourds.

Gout. Condition that manifests itself as recurrent attacks of acute Arthritis, distinguished from other forms of arthritis by the presence of increased uric acid in the body. It may become chronic and deforming. Gout usually begins with an acute attack of pain, inflammation, extreme tenderness, and redness in the affected joint. Treatment includes high liquid intake and drugs that increase uric acid excretion by the kidneys.

Grade. A unit of plane angle equal to one hundredth of a right angle. It is equivalent to 0.9 degree. The symbol is a superscript g used after the value; $1^g = \pi/200$ radian.

Gradient. (of a scalar). The product ΔA, where Δ is the operator del and *A* is a scalar. It is usually written grad *A*.

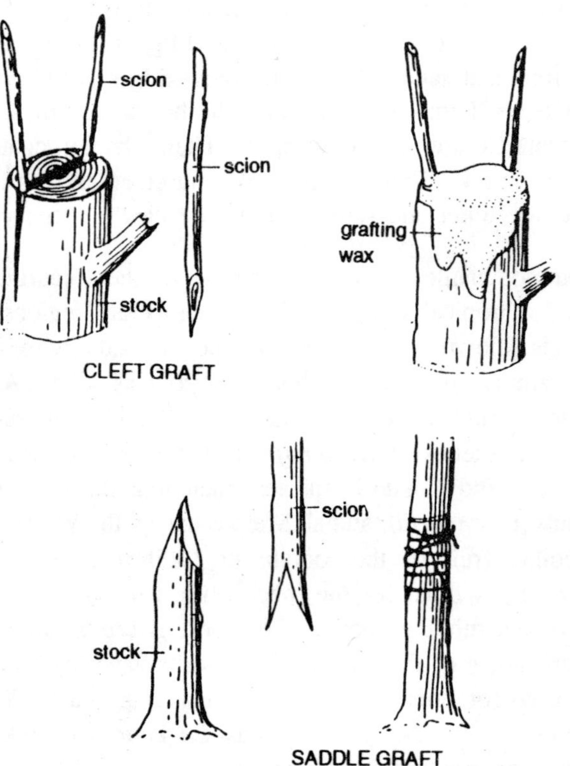

Fig. G-8. Two methods of grafting.

Grafting. Horticultural practice of uniting the cambial layers, or Cambium, of two closely related plants so that they grow as one. The scion (the piece grafted onto the stock or rooted part) may be a single bud or a cutting with several buds; the stock may be a whole plant or a root. The primary reason for grafting is to propagate Hybrids that do not bear seed or do not grow true from seed. It is also used to increase fruit tree productivity by adding to the numbers of buds, to grow a plant in an unfamiliar environment by using a stock adapted to that environment, and to combat diseases and pests by using resistant stock.

Graham's Law. The principle that the rate of diffusion of a gas is inversely proportional to the square root of its density. [After Thomas Graham (1805-69), British chemist.]

Graham, Thomas. 1805-69, Scottish chemist. His research in diffusion in both gases and liquids led to his formulation of Graham's law. The first to work in colloidal chemistry, he observed that some substances pass through a membrane more slowly that others, thus discovering Dialysis. His studies of phosphoric acid led to the present chemical concept of polybasic acids.

Grain. In agriculture, the caryopsis, or dry fruit, of a cereal Grass, the seedlike fruits of Buckwheat and other plants, and the plants bearing such fruits. Grains, whole or ground into meal or flour, are the main food of humans and domestic animals. The food content is mostly carbohydrate, but some protein, oil, and vitamins are present. Low in water content, grains can be stored for long periods. The primary grain crops—including Wheat, Rice, and Corn—together occupy about half of total world cropland.

Gram-atom. A unit of mass of an element equal to the element's atomic weight expressed in grams.

Gram-equivalent. A unit of mass equal to the equivalent weight of an element or group expressed in grams.

Gram-force. Symbol: gf. *See* gram-weight.

Gram (gramme). Symbol: g. A CGS unit of mass equal to one thousandth of a kilogram.

Gramme. *See* gram.

Gramme, Zenobe-Theophile. 1826-1901, Belgian electrical engineer. Model-maker for a Parisian manufacturer of electrical devices, he developed (1869), with little knowledge of electrical theory, an improved, practical direct-current dynamo. By reversing its principle, he invented the electric engine. In 1872, working with others, he transmitted direct current electricity over a long distance.

Gram-molecule. Symbol : gmol. A unit of mass equal to the molecular weight of a compound expressed in grams. *See* mole.

Gram's stain. Laboratory staining technique that distinguishes between two groups of bacteria, gram-negative and gram-positive, by identifying differences in their cell-wall structure. Pinpointing the type of bacterial pathogens involved in disease (see bacteria) can be valuable in determining treatment.

Gram-weight. Symbol : gwt. A unit of force equal to the force that could give a body of mass one gram an acceleration equal to the local acceleration of free fall. The force in dynes is the force in grams-weight multiplied by the local value of g in centimetres per second per second. The unit varies slightly from place to place. The *gram-force* (symbol: gf) is the force that would give a mass of one gram an acceleration equal to the standard acceleration of free fall (980.665 cm s^{-2}): 1 gf = 980.665 dynes.

Granite. Coarse-grained Rock composed chiefly of Quartz and Feldspars, of varying colour. It is commonly believed to be igneous (having solidified from a molten state), but some granites show evidence of being metamorphic in origin (see

Metamorphism). Formed a depth, granite masses are exposed at the earth's surface by crustal movements or erosion of overlying rocks. Very coarse-grained granite, called pegmatite, may contain minerals and gems of economic value.

Grape. Common name for the family Vitaceae, mostly climbing shrubs of tropical, subtropical, and temperate zones. The Wine grape *(Vitis vinifera)*, known since ancient times, probably originated around the Mediterranean and in W Asia, it is now grown in temperate regions of the Americas. Grapes grown east of the Rockies are usually hybrids of this species and disease-resistant American grapes, or varieties of native species. Raisins are dried grapes. The genus *Parthenocissus*, native to the U.S., includes the Virginia Creeper *(P. quinquefolia)* and Boston IVY *(P. tricuspidata)*.

Grapefruit. Evergreen tree *Citrus paradisi)* or the Rue family, and its globular Citrus Fruit, which grows in grapelike bunches and weighs from one to five pounds (0.45 to 2.27 kg). It is believed that the pomelo *(C. maxima)*, long a popular fruit in parts of Asia, was introduced into the West Indies, where a mutation produced the grapefruit.

Graph. Figure that shows relationships between quantities. In a bar graph the lengths of bars indicate the quantities to be compared. In a line graph the relationship to be studied is treated as a Function and is plotted on a coordinate system (for an example, see Cartesian Coordinates). The graph of a function $y = f(x)$, where x and y are numbers, is the set of points in the xy-plane with coordinates $[x, f(x)]$. Statistics makes use of both bar graphs and line graphs.

Graphite (black lead, Plumbago). A naturally occurring allotrope of carbon. It is a black solid material with a high melting point and a good electrical and thermal conductivity. The crystal structure has layers of covalently bonded hexagonally arranged atoms, with rather weak forces between the layers. This allows the slip of adjacent layers, a property that enables graphite to be used as a lubricant. It is also used in pencils, crucibles, electrodes, and as a moderator in reactors.

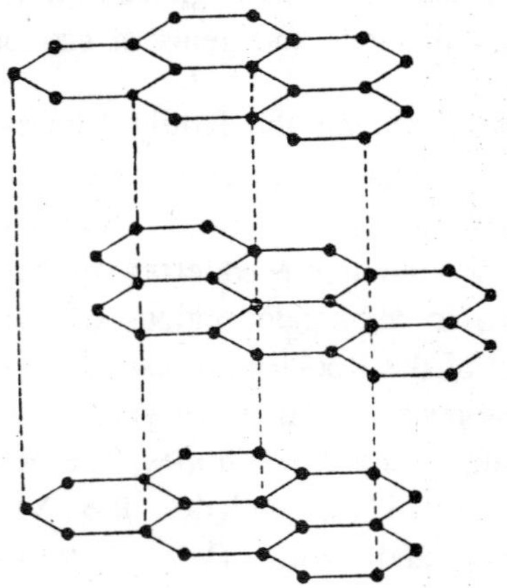

Fig. G-9. Graphite structure.

Graphite, plumbago, or **black lead.** Mineral, one of two crystalline forms of the element Carbon, occurring in various parts of the world. Dark gray or black, greasy, and soft, with a metallic luster, it is good conductor of electricity. It is used to make crucibles and electrodes; mixed with clay, it is the so-called lead of pencils.

Grass. Any plant of the family Gramineae, a widely distributed group of mostly annual and perennial herbs. The grass family is of far greater economic importance than any other. It includes the cereal grasses, e.g., Wheat, Rice, Corn, Oats, Barley, and Rye, which provide the Grain that is the staple food in most countries and the major type of feed; most hay and pasture plants, e.g., Sorghum and bluegrass; Sugarcane; and species of Reed and Bamboo used for thatching and construction. In addition, plants of the grass family provide raw material for the production of many liquors and Paper items, from climax vegetation (see Ecology) in areas of low rainfall (see Pampas; Prairie; Savanna; Steppe; Veld), and help to prevent Erosion.

Grasshopper. Slender, winged, singing Insect of two families in the order Orthoptera, having powerful hind legs, adapted for jumping and initiating flight, and chewing mouthparts. Usually only the males produce a song, but both sexes possess auditory organs. They range from $^1/_2$ to 4 in. (1-10 cm) in length. The long-horned grasshopper family includes the Katydid; the smaller, short-horned family includes the Locust.

Graticule. A rectangular network of fine wires or lines in the focal plane of an optical instrument; it is used as a reference system.

Grating. *See* diffraction grating.

Gravitation. The mutual attraction of bodies as a result of their mass. Gravitation is the effect causing objects to fall to the earth *(see* gravity). *Newton's law of gravitation* states that any two bodies each experience a force of attraction *(F)* proportional to the product of their masses m_1 and m_2) and inversely proportional to the square of their separation *(x)*: i.e. $F = Gm_1m_2/x^2$. G is the gravitational constant. Gravitational interaction is a much weaker force than electromagnetic interaction. A mass has an associated *Gravitational field* characterized by the fact that a force is exerted on a second mass placed in this field. According to the general theory of relativity, gravitation is explained by a curvature of space-time in the neighbourhood of matter.

Gravitation. The attractive Force existing between any two particles of matter. Because this force acts throughout the universe, it is often called universal gravitation. Issac Newton was the first to recognize that the force holding any object to the earth is the same as the force holding any object to the earth is the same as the force holding the moon and planets in their orbits. According to Newton's law of universal gravitation, the force between any two bodies is directly proportional to the product of their Masses and inversely proportional to the square of the distance between them. The constant of proportionality is known as the gravitational constant (symbol *G*) and equals 6.670 X 10^{-11} newton-M^2/kg^2 in the mks system of units. The measure of the force of gravitation on a given body on earth is the Weight of that body. In the general theory of Relativity, gravitation is explained geometrically; matter in its immediate neighbourhood causes the curvature of the four-dimensional Space-time continuum. See also Celestial Mechanics.

Gravitational Collapse. In astronomy, theoretically predicted final stage in the life history of a star (see Stellar Evolution). Gravitational collapse may begin when a star has depleted its steady sources of nuclear energy and can no longer produce the expansive force, which is a result of normal gas pressure, that supports the star against the compressive force of its own Gravitation. As the star shrinks in size (and increases in density), it may assume one of several forms depending upon its mass. A low-mass star may become a White Dwarf; an intermediate-mass star may explode a Supernova and become a Neutron star. If a star is too massive, however, its evolution will not stop at one of these stable configurations. According to present knowledge, nothing remains to prevent the star from collapsing without limit to an indefinitely small size. According to the general theory of Relativity, space becomes curved in the vicinity of matter; the greater the concentration of matter, the greater the curvature. When the radius of a star decreases below a certain limit (the Schwarzschild radius, or event horizon) determined by its mass, the extreme curvature of space seals off contact with the outside world. The former star is now a black hole. Because light and other forms of energy and matter are permanently trapped inside a black hole by the enormous pull of gravitation, a black hole can be observed only indirectly. For example, in some Binary-Star systems, such as the X-ray source Cygnus X-1, the smaller and invisible companions are strong emitters of X-rays and are suspected of being black holes. Gas flows from the larger component of the binary and forms an accretion disk around the black hole. As the gas from this disk spirals into the black hole, it is accelerated by the collapsed star's enormous gravitational pull; as this happens, the matter becomes compressed and heated to the extreme temperatures at which X rays are emitted.

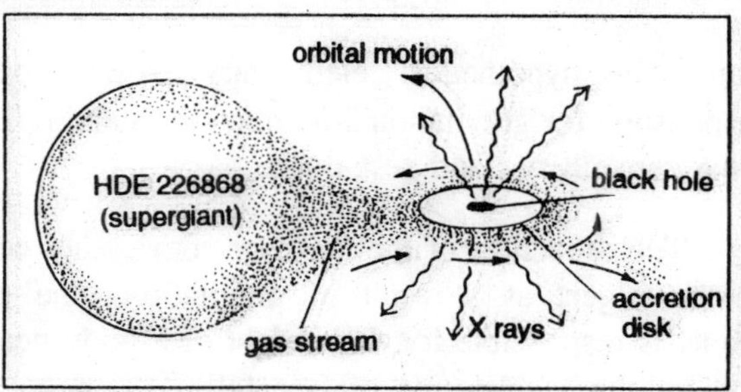

Fig. G-10. Gravitational collapse: The X-ray source Cygnus X-1 is believed to be a black hole (a massive star that has undergone gravitational collapse and cannot now be seen directly) orbiting around an optically observed supergiant star (HDE 226868).

Gravitational constant. Symbol: *G*. The constant appearing in the formula for Newton's law of gravitation. It has the value 6.664 X 10^{-11} newton metres-squared per kilogram squared (Nm^2 kg^2). *See also* Cavendish's experiment.

Gravitational field. *See* gravitation.

Gravitational interaction. A type of interaction occurring between bodies or particles as a result of their mass. It is about 10^{40} times weaker than the electromagnetic interaction. *See also* grativation.

Gravitational mass. *See* mass.

Gravitational waves. Waves propagated as a result of the acceleration or deformation of mass. The existence of gravitational radiation travelling at the speed of light is predicted by the general theory of relativity. In principle the radiation can be detected by its effect in exciting oscillations in a solid bar or similar detector. Usually experiments are performed using several detectors, widely separated, and observing coincidences between their signals. It has been suggested that gravitational radiation pulses are reaching earth form some point near the centre of the Galaxy, but the experimental evidence has been disputed.

Graviton. A hypothetical elementary particle postulated as responsible for gravitational interaction. Such a particle would have zero charge and rest mass.

Gravity. The attraction of the earth or some other celestial body for an object as a result of gravitation. The gravity of a planet is responsible for the weight of a body near its surface and for the acceleration of free fall.

Grease-spot photometer. A photometer in which a vertical sheet of white paper with a spot of grease in its centre is illuminated from both sides.

Greenockite. A naturally occurring from of cadmium sulphide, CdS.

Great circle. A circle on a sphere, whose centre is the centre of the sphere.

Greenhouse effect. Process whereby heat is trapped at the surface of the earth by the atmosphere. Energy form the sun passes through the atmosphere, warming the earth and providing the wavelengths used in Photosynthesis. Much of the incoming energy is reradiated in the form of heat, some directly and some as a result of either the Metabolism of living things or human industrial activities. This heat (infrared radiation) is prevented from leaving the earth by atmospheric carbon dioxide, water vapor, and ozone (acting like the glass in a greenhouse), and much of it goes back into the ground. An increase in atmospheric carbon dioxide of 10% over the past century has led some authorities to predict a long-term warming of the earth's climate.

Green Revolution. Popular term referring mainly to the large increases achieved in Grain production in certain underdeveloped areas — especially India, Pakistan, and the Philippines in the late 1960s. It was accomplished through the use of high-yielding Hybrids, chemical Fertilizers, and new crop strategies and harvesting methods.

Green vitriol. *See* iron sulphate.

Greenwich. Borough (1981 pop. 211,806) of Greater London, Se England, on the Thames R. The system of geographic longitude and time-keeping worked out at its old Royal observatory (1675-1958) have become standard in most of the world; the prime meridian (zero longitude) passes through the observatory. Other points of interest are the Royal Naval College (begun 17th cent.) and the National Maritime Museum.

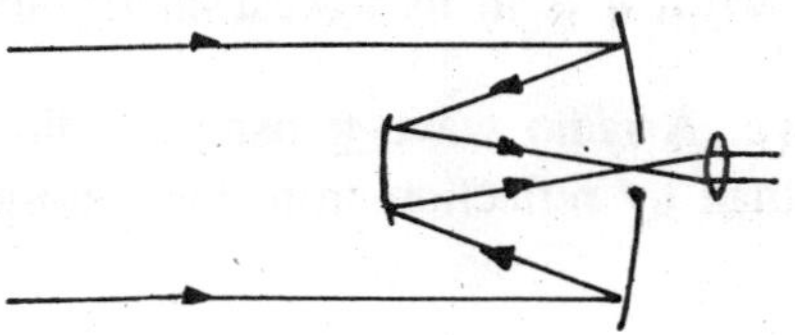

Fig. G-11. Gregorian telescope.

Gregorian telescope. A type of astronomical telescope similar to the Cassegrainian telescope but using a second concave mirror rather than a convex mirror (see illustration). [After James Gregory (1638-75). English scientist.]

Grenade. Small bomb designed to be thrown by hand or shot from a modified rifle or a grenade launcher. It may be filled with gas or chemicals but more often holds an explosive charge that fractures the casing into lethal fragments. First used in the 15th cent. (by "grenadiers"), the grenade later fell into disuse until the 20th cent., when it became a standard infantry weapon.

Grid. One of the electrodes controlling the current in a thermionic valve. The *grid bias* is the voltage between the grid and the cathode.

Grignard reagent. Any of class of organometallic magnesium compounds with the general formula RMgX, where R is an alkyl or aryl group and X is a halogen atom. The Grignard reagents are obtained by reacting alkyl or aryl halides with magnesium metal in dry ether. They are colourless crystalline solids, possibly having the form R_2Mg.- MgX_2. Grignard reagents are usually not extracted but are used in ether solution as intermediates in preparing many organic compounds. [After Francois Grignard (1871-1935), French chemist.]

Ground. *See* earth.

Ground state. The state of an atom, molecule, nucleus, or other system when it is in its lowest energy level.

Ground wave. A radio wave transmitted directly between points rather than by reflection from the ionosphere. *Compare* sky wave.

Group. 1. A group of atoms held together by covalent bonds in a chemical compound, considered as a unit of structure having characteristic properties. A *functional group* is the group in a molecule that is responsible for the compound's chemical behaviour. Thus acetaldehyde. CH_3CHO, contains the methyl group, CH_3^-, and the aldehyde group, -CHO. The aldehyde group is the functional group. A group is sometimes called a *radial,* although this term is now usually reserved for a free radical.

2. A set of elements belonging to the same vertical column in the periodic table.

3. A set of entities with certain mathematical properties. Members of the set can be combined together by addition, multiplication, or some other operation. They form a group if combination of any two members gives another member and obeys the associative law. Moreover there must be one member of the set (the *identity)* such that combination with this identity produces no change (as in multiplying a number by 1). Finally, for every member of the group there must be another member that combines with it to give the identity. The members of a group need not be numbers: they can be matrices, vectors, symmetry operations. etc. Group theory is used in spectroscopy, crystallography, and in particle physics (*see* unitary symmetry).

Grouper. Large, carnivorous Fish of the sea Bass family, abundant in tropical and subtropical seas and highly valued as food. Most have bright markings that change in colour and pattern to match the background. The largest is the spotted jewfish (up to 600 lb/270 kg). The red grouper and black grouper form the bulk of the commercial catch.

Group psychotherapy. A means of altering behaviour and emotions that is based on the use of group interactions. The technique was developed in the 1930s and received further impetus during World War II, when there was a great need for a prompt method of treating many patients simultaneously. The therapy can take many forms, depending upon the predisposition and training of the therapist, but is generally carried out in formal groups of 5 to 12 people; the members may be heterogeneous or homogeneous. The group becomes a "sample" social setting, reproducing conditions of interpersonal relationships in general; its members jointly participate in observing personal motivation and styles of interaction. Members can also attempt new behaviours and learn to deal with the consequences of these within he group. Group therapy may be combined with individual treatment.

Grouse. Henlike terrestrial Bird of the family Tetraonidae, found in the colder parts of the Northern Hemisphere. It is protectively plumaged in reds, browns, and grays. The American ruffed grouse *(Bonasa umbellus),* often miscalled Partridge or Pheasant, is a forest bird noted for the drumming sound made by the courting male. The ptarmigan *(Lagopus)* is an arctic species that migrates to the NW U.S. in winter, when its rusty brown plumage changes to white.

Guava. Small evergreen tree or shrub (genus *Psidium)* of the Myrtle family, native to tropical America and grown for its ornamental flowers and edible fruit. Species include the common tropical guava *(P. guajava)* and the strawberry guava *(P. cattleyanum)*. The guava is grown commercially in Florida and California and is made into jellies and beverages.

Guide dog. Dog trained to lead a blind person. Guide dogs were first trained to lead blinded veterans in post-World War I Germany. Schools to train such dogs now exist in several European countries and the U.S. Although the German Shepherd is the most widely used breed, other breeds, e.g., Boxer and Golden Retriever, are also employed.

Guinea pig. Domesticated form of the cavy *(Cavia porcellus),* a South American Rodent unrelated to the pig. Guinea pigs have rounded bodies and large heads, are 6 to 10 in. (15 to 25 cm) long, and weigh 1 to 2 lb (450 to 900 grams). There are short—and long-haired varieties and many colour combinations.

Gull. Aquatic Bird of the family Laridae, which also includes the Tern, found near all oceans and many inland waters. White with gray of black plumage, gulls are larger and bulkier than terms, and their tails are squared rather than forked. They have long, narrow wings adapted for soaring, webbed feet for swimming, and hooked bills.

Gum tree. Name for the Eucalyptus and for other trees, e.g., sweet gum (Witch-Hazel family) and Black Gum.

Gun. A device that discharges shot, shells, or bullets from a straight tube. A gun using the explosion of Gunpowder or some other explosive substance to propel the projectile is called a firearm; types include Artillery (large firearms), Mortar, and Small Arms. Certain other guns use compressed air produced by a spring-operated plunger (e.g., BB gun) or a lever-and-pump system (e.g., air rifle) to propel the projectile.

Gunmetal. A Bronze, an Alloy of Copper Tin, and a small amount of Zinc. Originally used to make guns, it is now employed in casting machine parts. The percentages of the three elements are varied, depending on the intended use.

Gunpowder. Explosive mixture of saltpeter (75%), sulfur (10%), and charcoal (15%). It is believed to have originated as early as the 9th cent. In China, where it was used for making fireworks, and to have been introduced into Europe in the 14th cent. Its use revolutionized warfare. Gunpowder was the only explosive in wide use until the mid-19th cent., when it was superseded by Nitroglycerin-based Explosives.

Gunpowder. An explosive mixture of potassium nitrate, sulphur, and powdered charcoal.

Gymnosperm. Plant in which the Seeds are exposed to the air during all stages of development, in contrast to Angiosperms. Gymnosperms, which include the pines, are woody plants with Stems, Roots and Leaves, and vascular, or conducting, tissue (xylem and phloem). Gymnosperms are always pollinated by wind, and many have their seed-bearing structures organized into cones. Constituting the plant division Pinophyta, they comprise the Cycads (subdivision Cycadicae); the Conifers, Yews, and Ginkgos (subdivision Pinicae); and the subdivision Gneticae, which contains three families in separate orders, all of evolutionary interest.

Gynecology. Branch of medicine specializing in disorders of the female Reproductive System. Modern gynecology deals with disorders of Menstruation, menopause, disease and maldevelopment of reproductive organs, hormonal and fertility problems, and Birth control devices. A related branch of medicine is obstetrics, which specializes in the treatment of women during Pregnancy and childbirth.

Fig. G-12. Three-frame gyroscope.

Gypsum. ($CaSO_4 2H_2O$), the most common sulfate mineral, occurring in many places and forms. It is very soft; the massive variety Alabaster and the lustrous variety satin spar are easily worked into decorative objects. Plaster of Paris, used to make casts, molds, and wallboard, is made from gypsum.

Gypsum. A naturally occurring form of hydrated calcium sulphate, $CaSO_4 2H_2O$, used in making plaster of Paris.

Gyro-compass. *See* compass.

Gyromagnetic ratio. Symbol : γ. The ratio of an atom, electron, nucleus, etc., to its angular momentum.

Gyroscope. A heavy wheel mounted so that it can spin freely at high speed and the axis of rotation can change its direction. Usually, a universal mounting is used so that the axile of the wheel can take any direction in space. If a torque is applied tending to change the direction of the axis. The gyroscope processes at a rate proportional to the size of the torque. In the diagram, the applied torque is produced by the weight of the gyroscope (i.e. by its tendency to fall over) and precession occurs about the vertical axis.

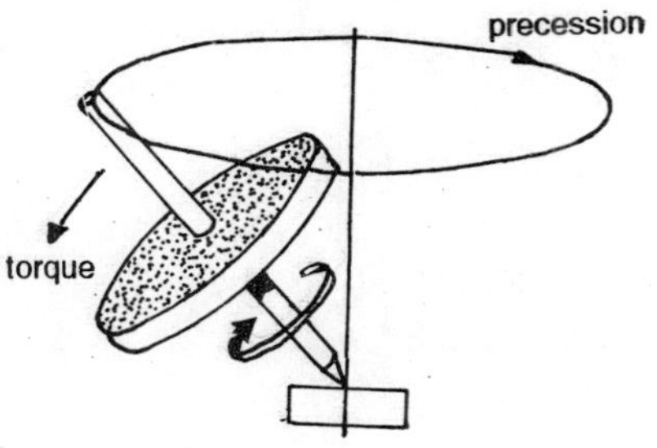

Fig. G-13. Gyroscope.

If the wheel has a large moment of inertia and the torque is small, its axis will preserve its direction in space. This effect is used in the gyro-compass, in stabilizers for ships, guidance systems for spacecraft, and similar devices.

Gyroscope. Symmetrical mass, usually a·wheel, mounted so that it can spin about an axis in any direction. A spinning gyroscope will resist changes in the orientation of its spin axis. Gyroscopes are used in ship stabilizers to counteracts rolling, and a gyroscope is the nucleus of most automatic steering systems, such as those used in airplanes, missiles, and torpedoes.

Haber, Fritz, 1868-1934. German chemist. During World War I he directed Germany's chemical warfare activities, which included the introduction of poision gas. After the Nazi rise to power (1933), he went into exile in England. For his discovery of the Haber process for synthesizing ammonia from its elements he won the 1918 Nobel Prize in Chemistry.

Haber process. An industrial process for manufacturing ammonia from nitrogen. A mixture of nitrogen and Hydrogen is passed over an iron catalyst at a temperature of 400-500° C and a pressure of 200 atmospheres. The equilibrium is $3H_3 + N_2 = 2NH_3$. The haber process is the most important method of nitrogen fixation for use in agricultural fertilizers. *See also* Le Chatalier's principle. [After Fritx Haber (1868-1934), German chemist].

Hadron. Any elementary particle that can take part in strong interactions. The hadrons include the baryons and the mesons: they exclude the photon and the leptons.

Haeckel, Ernst von, (he'kel) 1834-1919. German biologist and philosopher. He was an early exponent in Germany of Charles Darwin's theory of Evolution. Based on his interpretations of Darwin's theory, he evolved a mechanistic form of Monism. His theory of recapitulation postulated a hypothetical ancestral form represented by the gastrula stage of individual development. Although many were later proved incorrect, his theories attracted a following and stimulated research.

Haematite (bloodstone). A red or black naturally occurring form of iron III oxide, Fe_2O_3, used as the principal ore of iron and in paint pigments and jeweller's rouge.

Hafnium. Symbol: Hf. A highly lustrous silvery metallic element found in zirconium ores. It is usually obtained by the Kroll process or the van Arkel-de Boer process. Hafnium is used as a getter and as a neutron absorber in reactors. Its chemistry is similar to that of titanium and zirconium. A.N. 70; A.W. 178.49; m.pt. 2150°C; b.pt. 5400°C; r.d. 13.31; valency 2,3, or 4.

Hafnium. (Hf), metallic element, discovered by X-ray spectroscopy in 1923 by Dirk Coster and Georg von Hevesy. Lustrous, silvery, and ductile, it is chemically similar to Zirconium, the two elements are among the most difficult to separate. Hafnium is used for nuclear-reactor control rods.

Hail. Solid form of precipitation that can occur at any time of year, usually in a cumulonimbus cloud and often during a Thunderstorm. Large hailstones are spherical or irregularly spherical in shape and are composed of alternate hard and soft layers of ice. Hail usually forms when raindrops are blown up to high (cold) areas in a cloud and freeze. As they fall, they become coated with more water drops; they are then blown back up and refrozen, adding an additional layer. This process is repeated until the wind currents can no longer support the weight of the hailstone. Hailstones are generally less than 0.5 in. (1.3 cm) in diameter, although larger ones are occasionally observed.

Hair. Slender, threadlike outgrowth from the skin of mammals. Some animals grow a dense profusion of hair known as Fur or Wool. Insulation from cold, protection against dust and sand, and camouflage are among the functions of hair. Each hair originates in a deep, pouchlike hair follicle, which contains the bulb-shaped root of the hair. The papilla, a net of nerves and capillaries that supplies the hair, extends into an indentation at the base of the root. Here newly dividing cells force older cells upward, where they die and harden into the hair shaft. The shaft has two layers, the colorless cuticle and the cortex, which contains pigment and the protein keratin. Hair is lubricated by sebaceous glands in the follicle.

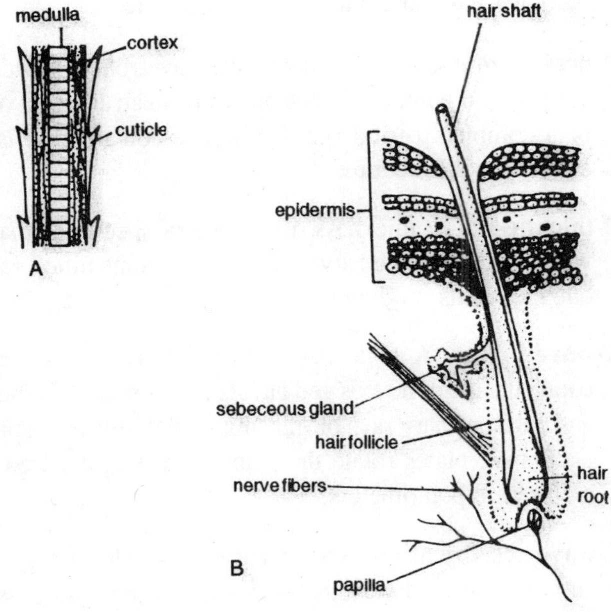

Fig. H-1. A Cross section of a hair shaft B. Central structure of a hair.

Hair hygrometer. A type of hygrometer in which the extension or contraction of a hair moves a pointer over a scale. The hair increases in length as the humidity increases and tension is maintained by a spring.

Half cell. An electrode dipping into a solution of an electrolyte. The simplest type consists of a metal electrode in contact with a solution of the metal ions of specified concentration. The electrode potential of such half cells is measured against a hydrogen electrode.

Half-life (half-value period). 1. Symbol: t. A measure of the stability of a radioactive substance, equal to the time taken for its activity to fall to one half of its original value. In this time one half of the atoms present disintegrate.

2. A similar measure of the rate of reaction or decay of elementary particles, excited atoms. Free radicals, etc.

Half-period zones. Zones into which a wavefront is divided so that at a forward point each zone differs from an adjacent one by half of a complete period (i.e. by l). The construction is used in explaining diffraction.

Half-thickness. The thickness of a piece of specified material that will reduce the intensity of a beam of transmitted radiation to one half of its original value.

Half-wave plate. A plate cut from a doubly refracting crystal parallel to the optic axis and having a thickness such that ordinary and extraordinary rays of transmitted light differ in phase by P. Half-wave plates rotate the plane of plane-polarized light and are used in polarimeters.

Half-wave rectification. Rectification of an alternating signal or current in such a way that current is allowed to pass only in alternate half cycles. *Compare* full-wave rectification.

Halide. Any of certain compounds that contain halogen atoms or ions. Such compounds are named *chlorides, bromides, icdides,* or *fluorides.* Simple ionic halides, such as sodium chloride, contain negative halide ions, Cl^-, Br^-, I^-, or F^-. For organic halides the names are usually used when the halogen is bound to a simple alkyl group, as in ethyl bromide, C_2H_5Br, or vinyl chloride, CH_2; CHCl.

Hall effect. An effect observed when an electric current is passed through a conductor under the influence of a transverse magnetic field. A potential difference is produced between the upper and lower surfaces of the conductor. This potential (the *Hall potential)* is caused by deflection of the moving charged particles. Its size depends on the concentration of charge carriers and its direction is determined by whether the carriers are electrons or positive holes.

Halley's comet. A comet with an orbital period of 76 years. One of the most spectacular and best-known comets, it last approached the vicinity of the sun in 1910. [After Edmund Halley (1662-1742), English astronomer who predicted its appearance in 1758-9].

Halley's comet. Periodic Comet named for Edmund Halley, who observed it in 1682 and identified it with those seen in 1531 and 1607. It returned in 1759, close to the time Halley predicted, as well as in 1835 and 1910. It was sighted again in late 1982 and will return to its closest approach to the sun early in 1986.

Hallucinogenic drug. Alkaloid substance that alters consciousness; also called psychotomimetic, or, popularly, psychedelic or mind-expanding drug. Hallucinogens include mescaline, or Peyote; psilocin and psilocybin, from the mushrooms *Psilocybe mexicana* and *Stropharia cubensis;* LSD (lysergic acid diethylamide); Belladonna; and Mandrake. Marijuana has hallucinogenic properties but is pharmacologically distinct. Hallucinogens have been used for centuries by primitive societies in both the Old and New Worlds to facilitate meditation, cure illness, placate evil spirits, and enhance mystical and magical powers. They produce a wide range of effects, from pleasant to extremely disturbing, depending on dosage, potency, and the personality and environment of the drug taker. Effects include altered perception of time and space and of the color, detail, and size of objects; also the experience of imaginary conversations, music, odors, tastes, and other sensations. Hallucinogens are not physically habit-forming, but tolerance—i.e., the need to take increased quantities to induce the original effect—may develop.

Halo. A luminous ring sometimes observed around the sun or moon. It is caused by refraction by small ice particles high in the earth's atmosphere.

Halo. In meteorology, circle of light surrounding the moon or sun. A halo occurs when sunlight or moonlight is refracted or reflected by ice crystals in the atmosphere, usually in a thin layer of high cirrostratus clouds. In general, a white halo results when the light is reflected, and colored rings (showing the colors of the spectrum, with red on the inside) result when the light is refracted.

Haloform. A compound with the formula CHX_3, where X is a halogen atom; i.e. fluoroform (CHF_3), chloroform ($CHCl_3$), bromoform ($CHBr_3$), and iodoform (CHI_3).

Halogen. Any of the five chemical elements in group Vlla of the Periodic Table. Fluorine, Chlorine, Bromine, Iodine, and the radioactive Astatine are nonmetallic, monovalent negative ions and exist in pure form as diatomic molecules. The first four elements exhibit an almost perfect gradation of physical properties. Fluorine is the least dense and chemically the most active, displacing other halogens from their compounds and oxygen from water, Iodine is the least active. The halogens form numerous compounds with each other, and with other elements, such as hydrogen halides, metal halides (salts), and halocarbons.

Halogenation. Any chemical reaction leading to the formation of a bond to a halogen atom, usually used in organic chemistry. Halogenation reactions, which can be either addition or substitution, include *fluorination* (F), *Chlorination* (Cl), and *bromination* (Br).

Halogens. The nonmetallic elements fluorine, chlorine, bromine, iodine, and astatine, belonging to group VIIA of the periodic table. They are monovalent and strongly electronegative, forming halides.

Hamilton, Alice. 1869-1970, American toxicologist; b. N.Y.C. On the faculty of Harvard Medical School (1919-35) and a pioneer in the study of industrial diseases and hygiene, she was sought out as an authority by political and governmental groups, including the state of illinois, the U.S. Dept. of Commerce, and the League of Nations health committee.

Hamilton, Sir William Rowan. 1805-65, Irish mathematician. A child prodigy, he had mastered 13 languages by the age of 13. Hamilton was one of the most creative mathematicians of his time. In *Theory of Systems of Rays* (1928) he unified the field of optics under the principle of varying action, which he later extended to dynamics and which is of fundamental importance in modern physics, particularly quantum theory. His later years were devoted to the development of his theory of quaternions.

Hamiltonian. Symbol : H.

A function used in mechanics to express the energy of a system. In simple causes it is the sum of the kinetic and potential energies. Thus for a particle of mass m and momentum p:$H = p^2/2m + V$, where V is the potential energy. [After Sir William Rowan Hamilton (1805-65), Irish mathematician and astronomer].

Hamster. Nocturnal Rodent with external cheek pouches for holding food. The common, or European, hamster (*Cricetus cricetus*), about 12 in. (30 cm) long, is reddish brown with white patches on the nose, cheeks, throat, and flanks. The Syrian, or golden, hamster *(Mesocricetus auratus)* of E Europe and W Asia is familiar as a laboratory animal and pet. Hamsters are serious agricultural pests in much of their range.

Hang gliding. Form of gliding in which a pilot jumps off an elevated point while suspended from a kitelike wing. The hang glider consists of a Rogallo wing about 18 ft (5.5 m) across and weighing about 35 lb (15.9 kg). Also called sky surfing, the sport began in California in the late 1960s.

Harcourt Pentane lamp. A lamp burning a mixture of pentane and air under specified conditions: used as a former standard for the international candle.

Hard. 1. Denoting electromagnetic or particle radiation with relatively high energy and, as a consequence, high penetrating or ionizing power Hard X rays are X-rays with short wavelength and high photon energy.

2. Denoting a vacuum in which there is a very low pressure of gas, such that ionization of gas plays little part in phenomena occurring in the vacuum.

Hardening. 1. The process of converting liquid vegetable oils, which contain unsaturated glycerides, into saturated compounds with higher melting points by catalytic hydrogenation. Usually nickel powder is the catalyst. The process is widely used in the manufacture of margarines.

2. Any process for making a metal harder. In general, a metal containing small amounts of another substance is less soft than the pure material because the foreign atoms in the lattice make it more difficult for slip of crystal planes to occur. Thus, iron is hardened by introducing carbon to produce steel.

Hardware. The equipment used in a computer, as distinguished from the programs—the *software.*

Hard water. Water that does not easily produce a lather with soap due to the presence of soluble calcium salts which react with the soap to form a scum of calcium salts of fatty acids. Two types of hardness are distinguished. *Temporary hardness* is removed by boiling and is caused by the presence of calcium hydrogen carbonate, formed by the action of dissolved carbon dioxide on calcium carbonate in limestone or chalk. Boiling decomposes the soluble hydrogen carbonate and precipitates calcium carbonate, which is insoluble. Softening can alsc be effected by adding clacium hydroxide to precipitate calcium carbonate (the *Clark process).* Temporary hardness is responsible for the formation of fur in kettles, boilers, and hot pipes.

Permanent hardness cannot be removed by boiling and is largely caused by calcium sulphate. It can be removed by sodium carbonate (washing soda), which precipitates calcium carbonate, or by use of Permutit or similar ion-exchange materials.

Hare. Herbivorous Mammal of the family Leporidae, which also includes the Rabbit, native to Eurasia, Africa, and North and Central America. The term *hare* is especially applied to the genus *Lepus,* sometimes called true hares. Hares generally have longer ears and hind legs than rabbits and move by jumping rather than running. They are usually brown or grayish in color, but northern species acquire a white coat in winter. Most North American hares are very large and are called jackrabbits.

Harmonic. A component of a complex wave with a frequency that is an integral multiple of the base frequency (the *fundamental).* The first harmonic has the frequency of the fundamental, the second harmonic (first overtone) has twice the frequency, etc. The complex wave is made up of the fundamental combined with the overtones.

Harmonic motion. Regular vibration, or back-and-forth motion, in which the acceleration of the vibrating object is directly proportional to the displacement of the object from its equilibrium position, but oppositely directed. A single object vibrating in this manner is said to exhibit simple harmonic motion; examples are a Pendulum swinging in a small arc, a mass bouncing at the end of a spring, and the motion of air molecules when a sound wave passes. Simple harmonic motion is periodic, i.e., it repeats itself at regular intervals. The time required for one complete vibration of the object is the period of the motion. The inverse of the period is the frequency, which is the number of vibrations per unit of time. The maximum displacement of the object from its central position of equalibrium is the amplitude of the motion.

Harmonic series. A mathematical series in which the reciprocals of each term form an arithmetic sequence. The simplest example is $1 + 1/2 + 1/3 + 1/4 + \dots$

Hartley. A unit of information equal to 3.321 93 bits ($\log_2$ 10 bits). It is used in digital computers. [After Sir Walter Noel Hartley (1846-1913), British physicist.]

Hartree. A unit of energy equal to e^2 / a_0, where e is the atomic unit of charge and a_0 is the atomic unit of length. It is equivalent to 4.8505 X 10^{-18} joule and is used in spectroscopy and other atomic studies.

Hawk. Name for smaller members of the Accipitridae family, diurnal Birds of prey, distinguished from the related Falcons by their broader, rounded wings. Hawks have keen sight, sharply hooked bills, and powerful feet with curved talons. The hunting hawks, or accipiters, include the goshawk, which feeds on small Mammals and birds, and the destructive chicken hawk. Buteos, or Buzzards, are a diverse group of larger hawks; they feed on Rodents and Reptiles. The term *hawk* is also applied to many falcons and a number of unrelated birds, e.g., the nighthawk (a Goatsucker), and certain Gulls and jaegers.

Hazel. Shrub or small tree (genus *Corylus)* of the Birch family, grown as an ornamental and for its edible nuts. Species include the American hazel *(C. americana)* and European hazels *(C. maxima* and *C. avellena);* hazelnuts, usually known as filberts, are chiefly from the European trees. Witch Hazel is not related to hazel.

Health physics. The study of the effects of radiation on health and of methods of protecting workers from radioactivity, X-rays, etc.

Heart. Muscular organ that pumps blood to all parts of the body. The pear-shaped human heart is about the size of a fist and lies just let of center within the chest cavity. The contractions of a heart muscle, or myocardium, are entirely self-stimulated. The heart is divided into two cavities by a wall of muscle; each cavity is divided in turn into two chambers, the upper ones called atria, the lower ones ventricles. Blood from the veins, high in carbon dioxide but low in oxygen, returns to the right atrium. It enters the right ventricle, which contracts, pumping the blood through the pulmonary artery to the Lungs. Blood rich in oxygen and poor in carbon dioxide returns from the lungs to the left atrium and enters the left ventricle, which contracts, forcing the blood into the aorta, from which it is distributed throughout the body. The blood is prevented from backing up by a series of valves.

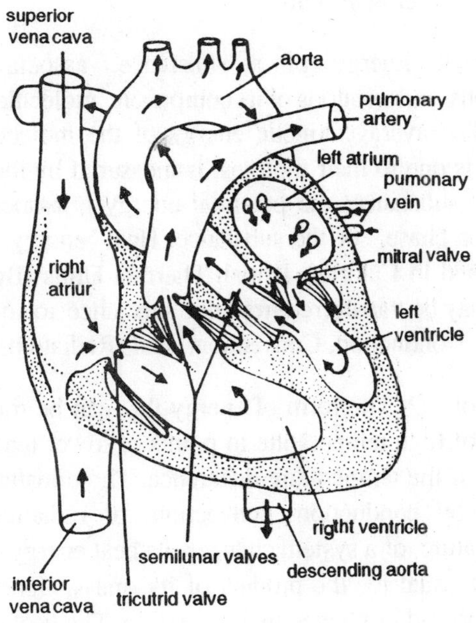

Fig. H-2. Cross section of the heart, with arrows indicating direction of blood flow.

Heart disease. Any of several abnormalities of the Heart and its function in maintaining blood circulation. Among the most common causes of heart disease are Birth Defects of the heart; infectious diseases, such as Rheumatic Fever; degenerative changes due to Coronary Artery Disease; and Hypertension. These and other forms of Cardiovascular disease, alone or in combination, can lead to congestive heart failure, a condition in which the heart cannot pump sufficient blood to meet the demands of the body. The various forms of heart disease may also cause disturbances in normal heartbeat, called arrhythmias.

Heat-lung machine. Device that maintains the circulation and oxygen content of the Blood when connected with the arteriovenous system. Used in open-heart surgery to bypass the Circulatory System of the Heart and Lungs, the oxygenator repeatedly draws off blood from the veins, reoxygenates it, and pumps it into the arterial system.

Heat. Internal Energy of a substance, associated with the positions and motions of its component molecules, atoms, and ions. The average kinetic energy of the molecules or atoms, which is due to their motions, is measured by the Temperature of the substance; the potential energy is associated with the state, or phase, of the substance. Heat energy is commonly expressed in Calories, British Thermal Units (Btu), or joules. Heat may be transferred from one substance to another by three means: Conduction, Convection, and Radiation.

Heat. Symbol : Q. The form of energy that can be transferred from points of high temperature to points of lower temperature as a result of the temperature difference. The transfer of heat can occur by conduction, convection, or radiation. When the temperature of a system changes, its heat energy changes by an amount equal to the product of its mass, its specific heat capacity, and its change in temperature. The heat of a system is the kinetic energy due to translation, rotation, and vibration of its molecules. Like other forms of energy, heat is measured in joules, although heat units such as calories and British thermal units are sometimes used.

Heat capacity (thermal capacity). Symbol C. The amount of heat required to raise the temperature of a body by one degree. It is usually measured in joules per kelvin.

Heat capacity or **Thermal capacity.** Ratio of the change in Heat energy of a unit mass of a substance to the change in Temperature of the substance. The heat capacity is a characteristic of a substance; it is often expressed in Calories per gram per degree Celsius of British Thermal Units per pound per degree Fahrenheit. See also Specific Heat.

Heat death. The state of a closed system when its total entropy has increased to its maximum value. Under these conditions there is no available energy; all the matter is disordered and the temperature is uniform. If the universe is a closed system it will eventually reach this state (the *heat death of the universe).*

Heat engine. A device in which heat energy is converted into mechanical energy.

Heater. A Filament used for heating the cathode in a thermionic valve.

Heat exchanger. Any device for transferring heat from one fluid to another. Heat exchangers are used for cooling, as in a car radiator, or for making use of waste heat, as in the outlet gases of furnaces, chemical plant, etc. They are made of metal and usually have coiled tubes, metal fins, etc., to increase the area and time of contact of the fluids.

Heat of atomization. The heat required to decompose one mole of a substance into its atoms.

heat of combustion. The heat evolved when one mole of a substance is burned in oxygen.

Heat of formation. The heat liberated or absorbed when one mole of a substance is formed from its elements.

Heat of neutralization. The heat evolved when one mole of an acid or base is fully neutralized. All strong acids and bases are fully ionized in solution and the neutralization is the reaction H^+ + OH^- = H_2O, which evolves 57 500 joules per mole.

Heat of reaction. The heat liberated or absorbed in a chemical reaction, usually expressed for molar amounts of reactants and products.

Heat of solution. The heat liberated or absorbed when one mole of substance is dissolved in a large volume of a specified solvent (i.e. enough to give effective infinite dilution).

Heaviside - Kennelley layer. *See* ionosphere. [After Oliver Heaviside 1850-1925) and Arthur Kennelley (1861-1939), English physicists.]

Heavy hydrogen. *See* deuterium.

Heavy water. A liquid with the same chemical properties as water but with a higher molecular weight and therefore different physical properties. The term usually applies to *deuterium oxide,* D_2O, in which the hydrogen isotope, deuterium, has an atomic weight of 2. Heavy water may be extracted from natural water by fractional distillation, exchange processes, or by electrolysis. It is used as a moderator in nuclear reactor. M.pt 3.8 C; b.pt. 101.4 C; r.d. 1.11 (25° C).

Hectare. A metric unit of area equal to 100 areas: i.e. 10,000 square metres or 2.471 05 acres.

Hedgehog. Usually brown or yellow Old World Mammal of the family Erinaceidae, related to Moles and Shrews. Spiny hedgehogs, of Africa and Eurasia, are covered with stiff spines; they can roll into a ball when attacked, becoming invulnerable to predators. The European hedgehog (*Erinaseus eropaeus*) is nocturnal, living in a burrow (often in hedgerows) and hibernating in winter. It is about 9 in. (23 cm) in length. Like a hog, it roots in the ground for food; it eats worms, insects, mice, frogs, and poisonous snakes.

Heimlich maneuver, emergency procedure used to treat choking victims whose airway is obstructed by food or another substance. It forces air from the lungs through the windpipe, pushing the obstruction out. If the victim is standing, the rescuer wraps his (or her) arms around the victim's waist; making a fist with one hand and placing the thumb side of the fist against the abdomen just above the navel, the rescuer grasps his fist with the other hand and presses in with firm, quick, upward thrusts.

Heisenberg uncertainty principle. The principle that it is impossible to measure both the momentum and position of a particle to any desired degree of accuracy: there is always an uncertainty Δp in the momentum and an uncertainty Δx in the position so that their product has the same order of magnitude as the Planck constant h: $\Delta p \ \Delta x \sim h$. This indeterminacy has nothing to do with the errors in the measuring instruments or with human errors of the observer. It arises from the fact that particles also have a wave-like behaviour (see de Broglie wave) and from the observation itself disturbing the system in an unpredictable way. A similar relationship holds for uncertainties in measurements of energy E and time t: $\Delta E \ \Delta t \sim h$. [After Werner Karl Heisenberg (b. 1901), German physicist.]

Heisenberg, Werner, 1901-76, German physicist. A founder of the Quantum Theory, he is famous for his uncertainty principle, which states that it is impossible to determine both the position and momentum of a subatomic particle (such as the electron) with arbitrarily high accuracy. The effect of this principle is to convert the laws of physics into statements about relative, instead of absolute, certainties. Heisenberg's matrix mechanics, a form of the quantum theory, was shown to be equivalent to Erwin Schrodinger's wave mechanics. Heisenberg received the 1932 Nobel Prize in physics for his work in nuclear physics and quantum theory.

Heliocentric. 1. Having the sun at the centre. A *heliocentric theory* of the solar system is one in which the planets are considered to move around the sun.

2. Measured with reference to the sun.

Helicopter. Type of aircraft in which the lift is obtained by means of one or more power-driven horizontal propellers called rotors. In a single-rotor helicopter the reaction torque from the spinning rotor is compensated by a small vertical rotor near the tail. On twin-rotor craft the rotors spin in opposite directions so that their reaction toques cancel each other. The helicopter is steered in any direction by inclining the axis of the main rotor in that direction. Because of its maneuverability and ability to land and take off in small areas, the helicopter is used in warfare and in a wide range of services, e.g., air-sea rescue, fire fighting, crop sowing, and traffic control. See Sikorsky, Igor.

Helios (he'leos). In Greek mythology, the sun god; son of the Titans Hyperion and Theia; father of Phethon. Each morning he left a palace in the east and crossed the sky in a golden chariot, then returned along the river Oceanus. He was a national god in Rhods, where a Colossus represented him. In Rome, where he was known as Sol, he was an important god.

Helium. Symbol: He. A colourless odourless tasteless gaseous element found in some natural gas, in radoactive ores such as pitchblende, and in the atmosphere, from which it can be obtained as a by-product in the liquefaction of air. It is used for filling airships and in some fluorescent lamps. Chemiscally, it is one of the inert gases and is not known to form any stable compounds. A.N. 2; A.W. 4.0026; m.pt. -272.2°C; b.p.t. -268 9°C;r.d.0.138 (air = 1).

Helium (He), gaseous element, first observed spectroscopically in the Sun during a solar eclipse in 1868. Its noncombustibility and buoyancy make this extremely unreactive, Inert Gas the most suitable of gases for Balloons and Airships. Deep-sea divers often breathe a helium-and-oxygen mixture; because helium is less soluble in human blood than nitrogen, its use reduces the risk of the bends (see Decompression Sickness; Diving, Deep-sea). Liquid helium is essential for low-temperature work (see Low-Temperature Physics; Superfluidity). Helium is also used in arc welding and gas-discharged lasers. Abundant in outer space, helium is the end product of fusion processes in Stars.

Helix. A curve in space that lies on the surface of a cone or cylinder and always makes a constant angle with the generators.

Helmholtz coils. A pair of identical flat coils of wire mounted parallel to each other with a separation equal to the coils' radius. The coils are connected in series and a uniform magnetic field is produced between them when a current is passed. [After Herman Ludwig Ferdinand von Helmholtz (1821-94), German scientist.]

Helmholtz free energy (Helmholtz function). Symbol: **A.** A free-energy function of a thermodynamic system equal to the internal energy minus the product of the temperature and the entropy: i.e. $A = U - TS$. [After Hermann Ludwig Ferdinand van Helmholtz (1821-94), German physicist.]

Helmholtz, Hermann Ludwig Ferdinand von, 1821-94. German scientist. An investigator of thermodynamics, electrodynamics, and vortex motion in fluids, he extended the application of the law of conservation of energy and in 1847 formulated it mathematically. A pioneer in physiological optics, Helmholtz invented (1851) the ophthalmoscope. He was professor of physics at the Univ. of Berlin.

Helmont, Jan Baptista van, 1577-1644, Flemish physician,chemist, and physicist. He attributes physiological changes to chemical causes, but his work was coloured by his belief in mysticism. He introduced the term gas in its present scientific sense and discovered carbon dioxide.

Hemiacetal. *See* acetal.

Hemihydrate. A solid compound with one molecule of water of crystallization of every two molecules of compound, as in calcium sulphate hemihydrate (plaster of Paris), $2CaSO_4.H_2O$.

Hemimorphite. A white or yellow mineral consisting of a hydrated zinc silicate, $Zn_4Si_2O_7(OH)_2.H_2O$, used as an ore of zinc.

Hemlock. Coniferous evergreen trees (genus *Tsuga*) of the Pine family, native to North America and Asia. The common hemlock of E North America (*T. canadensis*) has small cones and short, dark-green leaves that give the branchlets a flat appearance. It is grown as an ornamental and has been valued as a source of tanbark. The wood of a Western hemlock (*T. heterophylla*) is used for construction. Position Hemlock is an herb of the Carrot family.

Hamoglobin. Respiratory pigment found in the red blood cells of all vertebrates and some invertebrates (see Blood). It is produced in the bone marrow and carries oxygen to the body from the Lungs. An inadequate amount of circulating hemoglobin results in a condition called Anemia.

Hamophilia. Hereditary disorder in which the clotting ability of the Blood is impaired and excessive bleeding results. The hemophiliac bleed uncontrollably after a small cut or bruise, after rigorous exertion, or even spontaneously, unable to form or maintain a sufficient blood clot and occasionally needing a Blood Transfusion containing the missing clotting factor. The disease is genetically transmitted through females but usually affects only males.

Hemorrhoids or **Piles.** Dilation of the veins in or about the anus, often producing itching, bleeding, and pain. A common disordr associated with conditions such as constipation, pregnancy, and diarrhea, hemorrhoids can be treated with warm baths and suppositories; in severe cases, surgery may be necessary.

Hemp. Annual herb (*Cannabis sativa*) native to Asia and widely cultivated in Europe. It is often grown for the fiber made from its stems and for the narcotic drugs hashish and Marijuna, made primarily from the female flowers. Of major importance in many kinds of cord, hemp fiberis used in making paper, cloth, and other products. Hemp seed is used as bird food, and oil from the seeds is used in manufacture of paints, varnishes, and soap.

Henry. Symbol: H. An SI unit of inductance equal to the inductance of a closed circuit with a magnetic flux of one weber per ampere of current. [After William Henry (1774-1836), English chemist.]

Henry's law. The principle that the amount of gas dissolved in a certain volume of liquid is directly proportional to the equilibrium pressure of gas above the liquid. The law does not hold if the gas reacts with the solvent.

Henry, Joseph, 1797-1878. American physicist; b. Albany, N.Y. A professor philosophy at Princeton (then the College of New Jersey), he was later (from 1846) the first secretary and director of the Smithsonian Institution. Henry improved the electromagnet, invented and operated the first electromagnetic telegraph, and discovered self-inductance. The unit of inductance is often called the *henry* in his honor. Independently of Michael Faraday he discovered the principle of the induced current.

Hepatitis. Viral inflammation of the liver that causes nausea, fever, weakness, loss of appetite, and, usually, Jaundice. Two forms are most common: hepatitis A (infectious), spread through contaminated food or water; and hepatitis B (serum), usually transmitted by transfusions of infected blood or use of poorly sterilized medical instruments. Hepatitis can also occur as a complication of other diseases or as a toxic reaction to alcohol, drugs, or other chemicals. A vaccination against hepatitis B was approved for use in the U.S. in 1982.

Heptagon. A polygon that has seven sides.

Heptahydrate. A solid compound with seven molecules of water of crystallization per molecule of compound.

Heptane. A colourless flammable straight-chain liquid alkane, C_7H_{16} obtained from petroleum. Heptane is also called n-*heptane* to distinguish it from its branched chain isomers. It is used as a standard in determining octane rating, as a solvent, and in organic synthesis. M.pt. -906° C; b.pt. 98.4°C; r.d. 0.68.

Heptavalent (septivalent). Having a valency of seven.

Heptode. A type of themionic valve with seven electrodes (anode, cathode, and five grids).

Herb. Name for any plant that is used medicinally or as a Spice, and for the useful product of such a plant. The term is also applied to all herbaceous plants as distinguished from woody plants.

Herbicide. Substance that kills plants or inhibits their growth. Selective herbicides attack weeds without permanently harming crops; nonselective ones, generally toxic, are used to clear all plants from a broad area. Inorganic compounds such as common salts have long been used as herbicides; c.1900 certain sulfates, ammonium and potassium salts, and other compounds began to be used as selective herbicides. The 1940s saw the development of 2,4-D (2,4-trichlorophenoxyacetic acid), an organic compound that is a highly selective systemic herbicide. Such herbicides are now widely used. One such compound, 2,4,5-T (2,4,5-trichlorophenoxyacetic acid), was partially banned (1979) by the Environmental Protection Agency as potentially dangerous. 2,4,5-T was used in Agent Orange (a defoliant employed by U.S. forces in Vietnam) and is regarded by some as a cuase of illnesses suffered by veterans and their children.

Hermaphrodite. Animal or plant with both female and male reproductive systems, producing both eggs and sperm. Some hemaphrodites undergo self-Fertilization, in which egg and sperm of the same individual fuse, as in self-pollinating plants; others may undergo cross-fertilization, in which egg and sperm of different individuals fuse, as in the earthworm. Protandric hermaphrodites (e.g., oyster and sage plant) produce both eggs and sperm but at differnt times.

Hermit crab. Primarily marine animal (a Crustacean) distinguished from the true Crab by its long, soft, spirally coiled abdomen terminating in an asymetrically hooked tail. Most occupy the empty shells of Gastropods and are marine scavengers. The largest species reaches 1 ft (30 cm) in length.

Hernai. Protrusion of an internal organ or part of an organ through the wall of a body cavity. It may be present at birth or acquired after heavy strain on the musculature in such areas as the lower abdomen (inguinal hernia), diaphragm (hiatus hernia), or region around the navel. Surgery is sometimes recommended to alleviate symptoms and prevent the hernia from becoming caught and strangulated (cut off from the blood supply).

Heroin. Narcotic drug synthesized from Morphine. In many parts of the world it is accepted as an analgesic (for relief of pain), particularly for the terminally ill. In the U.S., it manufacture and importation are restricted, and its use in medicine remains under investigation. Heroin predominates in illicit narcotic traffic.

Heron. Bird of the family Ardeidae, large wading birds including the Bittern and Egret, found in many temperate regions but most numerous in tropical and subtropical areas. Herons have sharp, serrated bills, broad wings, and long legs. Their plumage is soft and drooping, and (especially at breeding time) they may have long, showy plumes on their heads, breasts, and backs.

Herpes simplex. Virus causing infection of the skin characterized by one or more blisters filled with clear fluid which usually appears around the lips (cold sores) or on the genitals. The initial infection, probably occuring during infancy or childhood, subsequently becomes dormant. The reappearance of blisters may be triggered by such factors as fever, exposure to sunlight, menstruation, or pregnancy. Genital herpes, a type of Venereal Disease, can be treated with the drug acyclovir.

Herring. Important marine or freshwater food Fish of the family Clupidae, including the sardine, menhaden, and Shad. Herrings are relatively small but very abudant; they swim in large schools. The adult common herring (*Clupea harengus*) of the N Atlantic is about 1 ft (30 cm) long, with silvery sides and blue back. The menhaden, of the Atlantic coast of North America, is a commercially important source of oil and fish meal. Of the smaller herrings and related species, the anchovies and sardines are the most important.

Herschel. Family of distinguished English astronomers. *Sir William Herschel,* 1738-1822, orginally Friedrich Wilhelm Herschel, b. Germany, discovered (1781) the planet Uranus, which led (1782) to his position as private astronomer to the king. The large reflecting telescopes that he constructed, including one with a 40-ft (12.2-m) focal length, far surpassed in size those of his contemporaries. He concluded from the motion of double stars that they are held together by gravitation and that they revolve around a common center, thus confirming the universal nature of Issac Newton's gravitational theory. He discovered the Saturnian Satellites Mimas and Enceladus (1789) and the Uranian satellites Titania and Oberon (1787). His research on nebulae suggested a possible origin of new worlds from gaseous matter, and his catalog of nebulae (including some objects that are now known to be galaxies or star clusters) increased those known from about 100 to 2,500. His sister, *Caroline Lucretia Herschel,* 1750-1848, discovered eight comets and three nubulae. For her arrangement of her brother's catalog of star clusters and nubulae, she received (1828) the Royal Astronomical Society's gold medal. Sir William's son, *Sir John Frederick William Herschel,* 1792-1871, spent four years (1834-38) at the Cape of Good Hope, cataloging 1,707 nebulae and clusters and 2,102 pairs of double stars. He published (1864) a consolidated catalog of 5,079 nebulae and clusters; revised by Johann Dreyer as *A New General Catalogue of Nebulae and Star Clusters* (NGC; 1888), it is still a standard reference. In photography, he was the first to use sodium thiosulfate as a fixing agent, and he introduced the terms *positive image* and *negative image.*

Herschelian telescope. A type of astronomical telescope in which light is reflected by a concave mirror at a small angle to the direction of the light.

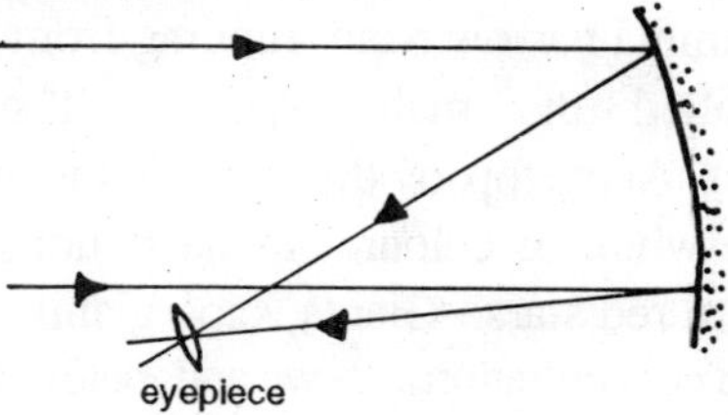

Fig. H-3. Herschelian telescope.

Hertz. Symbol: Hz. A unit of frequency equal to the frequency of a periodic process with a period of one second. The unit replaces the cycle per second, especially for expressing the frequency of waves or vibrations.

Hertzian waves. *See* radio waves.

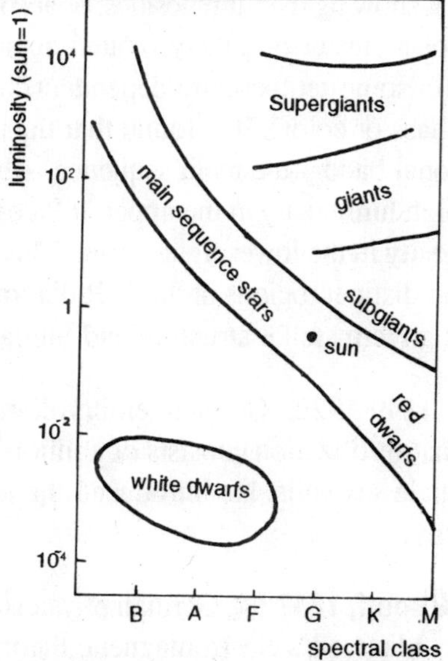

Fig H-4. Hertzspung-Russel diagram.

Hertzsprung-Russell diagram. 1. A graph in which the absolute brightness of stars is plotted against their spectral classification according to a system of spectral analysis that reflects decreasing surface temperature. Various classes of strars for concentrations on different parts of the graph. The greatest number of observed stars fall within a narrow band running from top left to bottom right and called the main sequence. The sun is a main-sequence star. At the top of this band lie the brightest and hotter stars, bluish-white in colour. At the bottom lie the cool and relatively dim red stars. Giants, supergiants and white drawfs form isolated concentrations above and below the main sequence. For the much older stars of the centre of the Galazy and of flobular clusters, a second Hertzsprung-Russell diagram may be drawn plotting brightness against colour. The diagram is important as a schematic indication of the evolution of stars. [Named after Ejnar Hertzsprung (1873-1969), Danish astronomer, and the American Henry Norris Russel (1879-1957).]

2. [For astronomers Ejnar Hertzsprung and Henry Norris Russell], graph showing the Luminosities, or absolute Magnitudes, of the Stars of a cluster or galaxy plotted against their surface temperature (or some temperature-dependent characteristic such as Spectral Class or color). It is found that the majority of stars lie on a diagonal band—the main sequence—that extends from hot stars of high luminosity in the upper left corner to cool stars of low luminosity in the lower right corner. The concentration of stars in certain distinct regions of the H-R diagram indicates that definite laws govern stellar structure and Stellar Evolution.

Hertwig, Oscar, 1849-1922. German embryologist. In 1875 he established that fertilization consists of a union of the nuclei of male and female sex cells. He introduced the term *coelom* for the body cavity.

Hertz, Heinrich Rudolf, 1857-94. German physicist. He confirmed James Clerk Maxwell's electromagnetic theory and produced and studied electromagnetic waves (radio waves), which he showed are long transverse waves that travel at the speed of light and can be reflected, refracted, and polarized like light. The unit of frequency, the hertz, is named for him.

Hess's law. The principle that if a set of reactants can be converted into a set of chemical products by a series of stages, the algebraic sum of the heats of reaction of the individual stages is equal to the heat of reaction for direct conversion of reactants into products.

Heterocyclic. Denoting a chemical compound that has a cyclic structure containing two or more types of atom in the ring. Pyridine, for example, is a common heterocyclic compound, having a hexagonal ring composed of five carbon atoms and one nitrogen atom. The odd atom in the ring, such as the nitrogen atom in pyridine, is termed the *hetero atom. Compare* homocyclic.

Heterodyne. Denoting the combination of two radiofrequency signals to give a resultant signal with a frequency equal to the difference in the two radiofrequencies. The phenomenon is the counterpart of beats produced by sound waves. If the radio signals are close together, the heterodyne frequency is in the audiofrequency range. The heterodyne principle is used in radio reception. The received carrier-wave signal is combined with a fixed signal generated in the receiver. The resulting heterodyne signal, modulated in the same way as the carrier wave, is amplified.

Hetergenous. Not uniform in composition; involving or containing more than one phase.

Heterogenous catalysis. *See* catalysis.

Heterogeneous reactor. *See* fission reactor.

Heterolytic fission. The breaking of a chemical bond with formation of two oppositely charged ions, as in the reaction AB = $A^+ + B^-$. *Compare* homolytic fission.

Heteropolar bond. *See* electrovalent bond.

Heusler alloy. Any of a class of alloys of managanese, copper, and aluminium that are ferromagnetic in spite of the fact that they contain no iron, cobalt, or nickel.

Hexachlorocyclohexane. Any of nine isomeric compounds, $C_6H_6Cl_6$, prepared by the chlorination of benezene in ultraviolet light. The most important is the symmetric isomer *1,2,3,4,5,6-hexachlorobenezene (lindane),* a white powdery solid used as an insecticide.

Hexadecanoic acid. *See* palmitic acid.

Hexagon. A polygon that has six sides.

Hexagonal. *See* crystal system.

Hexahedron. A polyhedron with six faces. A *regular hexahedron* is a cube.

Hexahydrate. A solid compound with six molecules of water of crystalization per molecule of compound, as in iron III chloride hexahydrate. $FeCl_3.6H_2O$.

Hexane. A colourless flammable straight-chain liquid alkane, C_6H_{14}, obtained from petroleum. Hexane is also called n-*hexane* to distinguish it from its branched-chain isomers. It is used as a solvent. M.pt. -95°C; b.pt. 69° C; r.d. 0.66.

Hexanedioic acid. *See* adipic acid.

Hexose. A simple sugar that has six carbon atoms in its molecules.

Hexyl group. The monovalent group C_6H_{13}, existing in several isomeric forms.

HF. *See* high frequency.

Highest common factor (HCF). The largest number that exactly divides into each of a given set of numbers. For example, the HCF of 12, 9 and 15 is 3.

High frequency (HF). A frequency in the range 3 megahertz to 30 megahertz.

High-speed steel. Any steel that retains its hardness at high temperatures, thus being suitable for making high-speed cutting tools, which become heated in use. High-speed steels contain tungsten (10-18%) and smaller amounts of other metals, such as chromium, vanadium, and molybdenum.

Hilbert, David. 1862-1943, German mathematician. After work on the theory of invariants, he developed a new approach that foreshadowed 20th-cent. abstract algebra. Hilbert also made important contributions to algebraic number theory and to functional analysis, particularly in the area of integral equations. His presentation in a 1900 speech of a list of 23 unsolved problems of mathematics stimulated much investigation by 20th-cent. mathematicians. Hilbert taught (1895-1930) at Gottingen Univ., where his breadth of mind and personality attracted to him many students who became important mathematicians.

Hipparchus. Fl. 2d cent. B.C., Greek astronomer; b. Bithynia (present-day Turkey). Ptolemy's geocentric theory of the universe was based largely on the conclusions of Hipparchus. In Ptolemy's *Almagest,* Hiparchus is credited with discovering the Precession of the Equinoxes, the eccentricity of the sun's apparent orbit, and certain inequalities of the moons' motions. He also made the first known comprehensive chart of the heavens.

Hippocrates. c.460-c.370 B.C., Greek physician, recognized as the father of medicine. He is believed to have been born on the island of Cos, where he later practiced and taught. Hippocrates based medicine on objective observation and deductive reasoning. Although he accepted the belief that disease results from an imbalance of the four bodily humors, he maintained that the humors were glandular secretions and that outside forces invluenced the disturbance. He taught that medicine should build the patient's strentgh through diet and hygiene, resorting to more drastic treatment only when necessary. The *Hippocratic Oath,* an ethical code formulated in ancient Greece and still administered to medical graduates in many modern universities, cannot be directly credited to him, but it does represent his ideal and principles.

Hippopotamus. River-dwelling Mammal *(Hippopotamus amphibius)* of tropical Africa, related to the pig (see Swine). The male stands about 5 ft (160 cm) at the shoulder and weighs about 5 tons (4,500 kg). The broad, short-legged body has a thick brown or gray hide, and the eyes are near the top of the head so the animal can see when submerged Hippopotamuses live in small herds, feeding on equatic plants. Hunted for meat and hides, they are endangered.

Histamine. Organic compound derived from the amino acid histidine. Histamine is released from certain cells upon tissue injury or during the activity of certain Antibodies. It then tends to dilate capillaries and to make them more permeable, which leads to edema (tissue swelling) and hives, accompanied by severe itching. This reaction, common to many allergies, can often be controlled with Antihistamines.

Histology. Study of the groups of specialized cells called Tissues, found in most multicellular plants and animals. Histologists study tissue organization at all levels, from the whole organ to the molecular components of cells. Histological techniques include tissue culture, fixing and staining, light and electron microscopy, and X-ray diffraction.

Hog cholera. Infectious viral disease of swine; also called swine fever. Perhaps the most serious disease of hogs in North America, it is characterized by listlessness, temperature increase, muscle tremors, and convulsions. Mortality is high; recovered hogs are permanently stunted.

Hole. A vacancy in all energy level in one of the energy bands of a semiconductor.

Holistic Medicine. System of health care based on a concept of the "whole" person as one whose body, mind, spirit, and emotions are in balance with the environment. Stressing personal responsibility for health, the holistic approach includes conventional medicine and Chiropractic as well as various nontraditional methods of diagnosis and therapy, e.g., acupuncture, Biofeedback, faith healing, folk medicine, megavitamin therapy, Meditation, and Yoga. Surgery and prescription drugs are generally avoided; instead, patients are encouraged to establish self-regulated regimes to control such illness-related factors as poor diet, smoking, alcohol intake, and stress. Although not officially recognized by the American Medical Association, holistic techniques have been of increasing interest to medical pactitioners and patients in the U.S. since the 1970s.

Holly. Common name for trees and shrubs of the Aquifoliaceae family, widely distributed but most numerous in central and South America. Many are cultivated as ornamentals. The English holly (*Ilex aquifolium*) and the American holly (*I. opaca*) are popular for their hard white wood and decorative spiny leaves and red berries. Some hollies are sources of tea, such as Mate, and medicinal preparations.

Holmes, William Henry. 1846-1933, American geologist and archaeologist; b. Harrison co., Ohio. After creating seminal reports on Yellowstone Park (see National Parks) and the classic illustration in the Grand Canyon atlas, Holmes became a pioneer in Southwestern archaeology and the study of Indian art, and was chief (1902-9) of the Bureau of American Ethnology. His later works include *Handbook of Aboriginal American Antiquities* (1919). He was also internationally recognized for his work as curator (1910-20) and director (from 1920) of the National Gallery of Art.

Holmium. (Ho), metallic element, discovered independently by J.L. Soret and M. Delafontaine in 1878 and by Per Teodor Cleve in 1879. The soft, malleable, lustrous, silvery metal is in the Lanthanide Series. Holmium, its oxides, and its salts have no commercial uses.

Holmium. Symbol : Ho. A soft silvery meatllic element belonging to the lanthanide series. A.N. 67; A.W. 164.9303; m.pt. 1470°C; b. pt. 2720°C; r.d. 8.78; valency 3.

Holography. A technique for recording three-dimensional images on photographic film using laser light. A beam of light from a laser is split by using a semitransparent mirror. One beam of light passes directly through the mirror onto the photographic film. The other is reflected onto the object to be photographed so that it is diffracted onto the film. Interference of the beams produces a pattern of spots on the film (the *hologram*). If the film is them illuminated directly by a beam of the same laser light, it acts as a diffraction grating and gives a real image of the object and a three-dimensional virtual image.

Homocyclic. Denoting a chemical compound that has a cyclic structure in which the ring is composed of identical atoms. Benezene is a common example of a homocyclic compound. *Compare* heterocyclic.

Holography. Method of reproducing a three-dimensional image of an object by means of light-wave patterns recorded on a photographic plate or film. The object is illuminated with a coherent beam of light produced by a Laser. Before reaching the object, the beam is split into two parts: the reference beam is recorded directly on the photographic plate, and the other is reflected from the object and then is recorded. On the photographic plate the two beams create an Interference pattern, exposing the plate at points where they arrive in phase. When this photographic recording, called a hologram, is later illuminated with coherent light of the same frequency as that used to form it, a three-dimensional image of the object becomes visible, and the object can be photographed from various angels.

Homicide. Killing of one human being by another human being. In law, there are four categories of homicide. Two of them, murder and manslaughter, are criminal; both are homicide commited without justification or excuse, but manslaughter is distinguished from murder by the absence of malice aforethought. Noncriminal homicides are justifiable homicide, which is killing in circumstances authorized by law (e.g., executing a condemned prisoner), and accidental, negligent, or excusable homicide, which is killing unintentionally, by misadventure, or without gross negligence (e.g., in the course of unsuccessful surgery).

Home erectus. Early human species dating from 1,500,000 to 300,000 years ago. He had an upright posture; in body size and brain development he is intermediate between the African hominid *Australopithecus* and later species of *Homo sapiens*. His Culture included stone tools and the first use of fire. The first fossils, found (1891) in Java, were called *Pithecanthropus*, or Jawa Man. Others were *Sinanthropus*, or Peking man (China); Heidelberg man (Germany); and Chellean man (E. Africa).

Homogeneous. Uniform in composition; involving or containing a single phase.

Homogeneous catalysis. *See* catalysis.

Homogeneous reacter. *See* fission reactor.

Homologous series. A series of related chemical compounds for which the formula of each member differs from the preceding one by the same group of atoms, usually a CH_2 group. For example, the alcohol CH_3OH, C_2H_5OH, C_3H_7OH, etc., form a homologous series. Two successive members of such a series are called *homologues*.

Homolytic fission. The breaking of a chemical bond with formation of two neutral atoms or radicals, as in the reaction AB = A- + B- *Compare* heterolytic fission.

Homopolar bond. *See* covalent bond.

Homopolymer. *See* polymer.

Homoxexuality. Sexual interest in members of one's own sex, sometimes accompanied by fear of or aversion to sexual activity with the opposite sex. Female homosexuality is known as lesbianism. Medical and psychological research has yielded little evidence that homosexuality is caused by either biological predisposition or feelings that one belongs to the opposite sex. No theory is conclusive but many have been proposed, including the psychoanalytic theory attributing homosexuality to an inability to resolve issues arising during the Oedipal stage of development as well as the suggestion by some authorities that homosexuality is an expression of nonsexual problems, such as fear of adult responsibility. Many individuals experiment with homosexuality during preadolescence and Adolescence but grow up to assume the heterosexual role. Some authorities as well as homosexual groups believe that instance of mental disorders in homosexuals result more from societal pressures and bias against homosexuality than from any psychopathology. In recent time (1975) the American Psychiatric Association dropped the classification of homosexuality as a mental disorder, and groups such as Gay Liberation have formed to fight for the civil rights of homosexuals.

Honey. Sweet Viscous fluid made by honeybees from the nectar of flowers, containing 70% to 80% sugar. It has been a major sweetening agent since earliest times. The worker bee transforms the sucrose of nectar into the simple sugars fructose and glucose by the enzyme action of its honey sac, and stores the honey to thicken in the wax cells of the hive. The excess of the colony's requirement may be extracted by humans for use as food. The flavor and color of honey depends on the kind of flower that produced the nectar, e.g., alfalfa, clover, orange, or tupelo.

Honey locust. Deciduous tree (*Gleditsia triacanthos*) of the Pulse family, native to the eastern half of the U.S. Often grown as a shade tree or as an ornamental, the honey locust has fragrant flowers, compound leaves, branching thorns, and brown pods with edible pulp. Its durable wood is used chiefly for fence posts and crossties.

Honeysuckle. Common name for vines and shrubs of the family Caprifoliaceae. Found in the Northern Hemisphere, the family includes the honeysuckles, elders, snowberries, viburnums, and weigelas. The true honeysuckles (genus *Lonicera*) include the well-known trumpet honeysuckle (*L. sempervirens*), with fragrant scarlet blossoms, and the Japanese honeysuckle (*L. japonica*), with white to yellow blossoms, now a naxious weed. Elders and viburnums usually have showy clusters of white flowers, and some produce edible berries, e.g, the North American elder (*Sambucus canadensis*), whose berries are used in preserves and wine.

Hoof-and-mouth disease. Highly contagious viral disease of cattle, sheep, goats, and other cloven-hoofed animals. symptoms include fever, loss of appetite and weight, and blistering on mucous membrane. The disease is spread through direct contact and also indirectly by way of contaminated food. Water, soil, and other materials. Humans seldom contract the disease but may be carriers, as may rats, dogs, birds, and wild animals. Although there are vaccines for the disease, epidemics still cause enormous losses to the livestock industry.

Hooke's law. The principle that the strain produced in an elastic material is proportional to the stress producing it. The law holds only up to the material's elastic limit. *See also* elasticity. [After Robert Hooke (1635-1703). English physicist.]

Hooke, Robert, 1635-1703. English physicist, mathematician, and inventor; considered the greatest mechanic of his age. He improved astronomical instruments, watches, and clocks, and first formulated the theory of planetary movements as a mechanical problem. He devised (1684) a practicable telegraph system; invented the spiral spring in watches and the first screw-divided quadrant; and constructed the first arithmetical machine and Gregorian telescope. He stated Hooke's law on elasticity and anticipated the law of universal Gravitation.

Hooker, Sir William Jackson, 1785-1865. English botanist. A leading authority of ferns, he formed a famour herbarium and built up the Glasgow and Kew botanical gardens. In Kew Gardens he founded the first museum of economic botany. His many writings include *British Jungermanniae* (1816), *Flora Scotica* (1821), British Flora (1830), and works on ferns.

Hookworm. Any of a number of parasitic nematodes, order Strongiloidae, found in tropical and subtropical climates. The hookworm larva usually penetrates exposed skin of humans and other mammals and migrates to the small intestine, where it attaches itself by means of hooks and feeds on the host's blood. Hookworm infestation causes Anemia, diarrhea, and abdominal pain, and is treated with drugs.

Hop. Herbaceious perennial vine of the Mulberry family, widely cultivated since early times for brewing pruposes. The commercial hop (*Humulus lupulus*), native to Eurasia, is grown for the conelike female flowers, called hops, used to impart a bitter flavour to beer. Oil of hops is used in perfumes, and the stem is used for fiber.

Horizontal component. B_0. The component of the earth's magnetic field in a horizontal direction at a point on the earth.

Hormone. Chemical messenger released in minute amounts by the endocrine, or ductless, glands and carried by the bloodstream to target tissues, where it produces either rapid or long-term effects. Most hormones fall into two major categories: Peptides (chains of amino acids) and Lipids (including Steroids). Since the lack of any hormone may cause serious disorders, many hormones are now synthesized for use in treating such hormonal deficiencies.

Horned lizard or **Horned toad.** Broad, flat-bodied Lizard (genus *Phrynosoma*) found in arid regions from SW Canada to Guatemala. The body is 3 to 5 in. (7.6 to 12.7 cm) long, with spines on the head, sides, and back, Horned lizards are protectively coloured, usually in dull grays or browns.

Horn Silver. A naturally occuring form of silver chloride, AgCl, used as an ore of silver.

Horse. Hoofed, herbivorous Mammal of the family Equidae, the single living genus of which is *Equus*, which includes the domestic horse, the wild Przewalski's horse, the Ass, and the Zebra. All are swift, plains-dwelling herd animals with teeth adapted for grinding coarse grass. The species can interbreed (see Mule), but offspring are usually sterile. The modern horse evolved in North America (where it later became extinct) and spread over the world. It was hunted by early man and domensticated by Asian nomads in the 3d millennium B.C. The fast, light southern breeds may have originated independently of the heavier northern draft breeds, and the small breeds called ponies may have evolved from a wild European race. Until recently horses were widely used for warfare, agriculture, and transportation. Today light breeds are used for show and sport.

Horse chestnut. Common name for some trees and shrubs of the family Hoppocastanaceae, found in north temperate zones and South America. The horse chestnut tree (*Aesculus hippocastanum*) is often cultivated for shade and ornament. The family includes the buckeyes; the prevalence of the Ohio buckeye (*A. glabra*) has given Ohio its nickname, the Buckeye State. The soft wood of horse chestnuts and buckeyes is used for paper pulp, carpentry, and woddenware.

Horsepower. Symbol: hp. An imperial unit of power equal to 550 foot-poundsforce per second. It is equivalent to 745.700 watts.

Horseeradish. Perennial herb (*Armoracia rusticana*) of the Mustard family, native to central and S. Europe. Once used medicinally against scurvy, today it is grown mainly for its pungent roots, which are used in a popular condiment.

Horsehoe crab. Large marine invertebrate animal (class Merostomata), an Arthropod. It has a heavy, dark brown domed exoskeleton divided into a broad horseshoe-shaped front part (prosoma), a tapered middle region (opisthosoma), and a spiky, taillike part (telson). Also known as king crabs, horseshoe crabs swim or burrow using five pairs of walking legs. The largest reach 2 ft (61 cm) in length.

Horsetail. Plant of the genus *Equistum*, the single surviving genus of a large group of primitive vascular plants that flourished during the Carboniferous period. Found in temperate and tropical regions, horsetails seldom grow to over 3 ft (91 cm), although Fossil evidence indicates that many extinct species were treelike in size. They have whorls of small, scalelike leaves around a green, hollow, jointed stem. Some species bear shoots with spore-bearing cones at the top for reproduction. The scouring rush (*E. hymenale*), common in North America and Eurasia, is so called because it has a coarse texture suitable for scouring.

Horticultre. Science of cultivating fruits, vegetables, flowers, and ornamental plants; a branch of Agriculture. Horticulture usually refers to small-scale gardening and Agronomy to the large-scale growing of field crops. Although many horitcultural practices are ancient, relatively recent knowledge of Genetics, plant Physiology and Pathology, Biochemistry, Ecology, Entomology, and Soils, and the practical application (e.g., in plant breeding) of this knowledge, have made horticulture an extremely complex science.

Hot. 1. Denoting an object of substance that is radioactive.

2. Denoting an atom, ion, or molecule in an excited electronic state, especially one excited by a nuclear decay process.

Hot spring. Natural discahrge of groundwater with an elevated temperature. Most hot springs (including Geysers) result from water passing through or near recently formed, hot, igneous rocks. Current energy concerns have spurred an interest in using the geothermal energy of hot springs.

Hot-wire ammeter. *See* ammeter.

Hudin, Jean Eugene Robert, 1805-71. French magician. Famed for his optical illusions, he liked to explain the natural causes of these and other "magic" tricks.

Hound. Class of dogs bred to hunt animals. Most hunt by scent, their quarry ranging from such large game as bear or elk to small game and vermin. Some long-legged breeds that hunt mainly by sight are also classed as hounds. A third variety, treeing hounds, track by scent and pursue tree-climbing animals. Many scent hounds have coats patterned in "hound colours": black, white and tan.

Hubble constant. Symbol: *H*. The ratio of the recessional velocity of a star, as measured by its red shift, to its distance. If it is a constant it is the age of the universe: 7.6 x 10^9 years. [After Edwin Hubble (1889-1953), American astronomer.]

Hubble, Edwin Powell, 1889-1953. American astronomer; b. Marshfield, Mo. As a staff member (from 1919) at Mt. Wilson Observatory, Hubble used the 100-in. (254-cm) telescope there to discover that there are large-scale galaxies beyond the Milky Way and that they are distributed almost uniformly in all directions. In what is now known as Hubble's Law, he was the first to offer observational evidence supporting the theory of the expanding universe.

Hubble's law. Statement that the greater the distance between any two Galaxies, the greater is their relative speed of separation. In other words, the universe is expanding roughly uniformly. The empirical finding is more consonant with the big bang theory of the universe's origin than with the steady state theory. The law was first proposed in 1929 by Edwin Hubble, who observed that the more distant a galaxy, the greater is its Red Shift and hence its velocity relative to our galaxy.

Huckel rule. *See* aromaticity.

Hue. The property of a colour determined by the wavelength of the light. The hue of a colour is the quality that allows an observer to distinguish it as resembling a spectral colour—a red hue, a violet hue, etc.

Huggins, Sir William, 1824-1910. English astronomer. A pioneer in spectroscopic photography, he helped develop the combined use of the telescope, spectroscope, and photographic negative. He adapted the gelatin dry-plate negative for use in astronomical photography, making possible exposures of any length. Huggins proved that while some nebulae are clusters of stars, others are uniformly gaseous. He made (1866) the first spectroscopic observations of a nova.

Humidity. A measure of the concentration of water vapour in the atmosphere. The *absolute humidity* is the mass of water vapour per unit volume of air. The *relative humidity* is the ratio of the mass of vapour per unit volume present in the air to the mass per unit volume that would be present if the air were saturated at the same temperature.

Humidity. Moisture content of the atmosphere, a maor element of Climate. Humidity measurements include *absolute humidity,* the mass of water vapor per unit volume of natural air; *relative humidity* (usually meant when the term *humidity* alone is used), the ratio of the actual water-vapor content of the air to its total capacity at a given temperature; and *specific humidity,* the mass of water vapor per unit mass of natural air. Relative fumidity is usually measured by means of a Hygrometer.

Hummingbird. Small, colorful Bird with a long, slender bill, of the New World family Trochilidae, found chiefly in the mountains of South America. Hummingbirds vary in size from the $2^1/_4$-in. (6-cm) fairy hummingbird of Cuba, the smallest of the birds, to the $8^1/_2$-in. (21.6-cm) giant hummer of the Andes. They are usually seen hovering or darting (at speeds of up to 60 mi/97 km per hr) in the air, beating their wings at 50 to 75 beats per sec.

Constant feeding supplies their enormous energy needs. At night they lapse into a state of torpor similar to Hibernation.

Hurricane. Tropical Cyclone formed over the North Atlantic Ocean in which the winds attain speeds greater than 75 mph (121 km/hr). A tropical cyclone passes through two stages, tropical depression and tropical storm, before reaching hurricane force. An average of 3.5 tropical storms per year become hurricanes; one to three of these approach the U.S. coast. Hurricanes usually develop between July and October. A hurricane is nearly circular in shape, and its winds cover an area about 500 mi (800 km) in diameter. As a result of the extremely low central air pressure (around 28.35 in./72 cm of mercury), air spirals inward toward the hurricane's eye, an almost calm area about 20 mi (30 km) in diameter. Hurricanes, which may last from 1 to 30 days, usually move westward in their early stages and then curve northward toward the pole. Deriving their energy from warm tropical ocean water, hurricanes weaken after prolonged contact with colder northern ocean waters, becoming extratropical cyclones; they decay rapidly after moving over land areas. The high winds, coastal flooding, and torrential rains associated with a hurricane may cause enormous damage. Tropical cyclones that form over the Pacific Ocean are called typhoons.

Hutoon, James, 1726-97. Socttish geologist who formulated controversial theories of the origin of the earth (see Uniformatarianism) that paved the way to modern geology. His great work was *The Theory of the Earth* (2 vols., 1795). John Playfair, in his *illustrations of the Huttonian Theory of the Earth* (1802), simplified Hutton's theories.

Huxley, Thomas Henry, 1825-95. English biologist and educator. He gave up his own biological resarch to become an influential science publicist and was the principal exponent in England of Charles Darwin's theory of Evolution. An agnostic, he doubted all things not immediately open to logical analysis and scientific verification. However, he placed human ethics outside the scope of materialistic evolutionary processes, and believed that progress was achieved by the human control of evolution.

Huygens, Christiaan (ho'gans), 1629-95. Dutch mathematician and physicist; son of Constantijin Huygens. He improved telescopic lenses, was the first to interpret correctly the ring structure surrounding Saturn, and discovered its satellite Titan. He was the first to use a pendulum in clocks. **Huygens** developed a wave theogy of Light opposed to Isaac **Newton**'s corpuscular theory and formulated Huygens's principle of light waves, which holds that every point on a wave front is a source of new waves. He discovered the polarization of light in calcite.

Hyacinth. Bulbous herb (genus *Hyacinthus*) of the Lily family, native to the Mediterranean region and South Africa. The common hyacinth, cultivated primarily in Holland, has a single spike of fragrant flowers in shades of red, blue, white, or yellow. The smaller, related grape hyacinth (*Muscari*), mostly blue-flowered, is also commonly cultivated.

Hyatt, John Wesley, 1837-1920. American inventor; b. Starkey, N.Y. He is known especially for his development of Celluloid, which he manufactured with his brothers beginning in 1872. Among his other inventions were the Hyatt filter, a means of chemically purifying water while it is in motion, and a widely used type of roller bearing.

Hybrid computer. A type of computer in which a continuously variable input is converted into digital form for processing. Hybrid computers have characteristics of both analog and digital computers.

Hybrid orbital. An orbital formed by combination of two or more atomic orbitals of an atom. Carbon, for example, has one *s* orbital and three *p* orbitals. These may be thought of as mixing to give four equivalent sp^3 hybrids, each directed towards the corner of a tetrahedron. It is these that overlap with *s* orbitals of hydrogen in methane. Other types of *hybridization* can occur:

sp^2 hybrids are three coplanar orbitals at 120° to each other; *sp* hybrids are two colinear orbitals directed away from each other. The sp^2 hybrids occur in ethene and *sp* hybrids in acetylene.

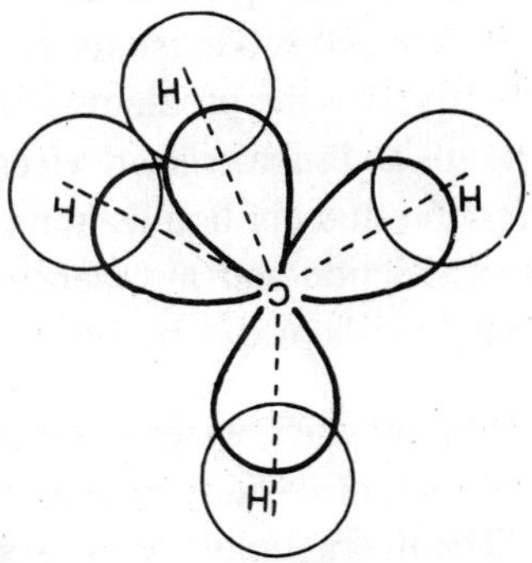

Fig. H-5. sp^3 hybrid orbitals in methane.

Hydra. Freshwater organism (class Hydrazoa) widely distributed in lakes, ponds, and sluggish streams; a Coelenterate. Hydras are small, cylindrical, solitary animals, about 1 in. (2.5 cm) long, that attach themselves temporarily to submerged objects by means of a disk at the anal end. Tentacles equipped with stinging cells (nematocysts) surround the mouth; hydras use these to stun their prey.

Hydraulic machinery. Machines that derive their power from the motion or pressure of water or some other liquid. Water or oil under pressure is commonly used as a source of power for many types of machines. The hydraulic press, used, e.g., to form three-dimensional objectsfrom sheet metal or plastic and to compress large objects, consists of two cylinders of different size, each filled with liquid and fitted with a piston, and each connected to a pipe filled with the same liquid. According to Pascal's law, pressure exerted upon the smaller piston is stransmitted undiminished through the liquid to the surface of the larger piston, which is forced upward. A small pressure exerted on the smaller piston creates a stronger force on the larger piston because the area of the latter is larger and the distance it moves is less. The same principle is used to power the hydraulic jack, which is used to lift heavy loads. The hydraulic Elevator is also an application of Pascal's law.

Hydraulics. Branch of engineering that studies the mechanical properties of fluids. There are two subdivisions. *Hydrostatics,* the study of liquids at rest, involves the problems of buoyancy and flotation, pressures on dams and submerged devices, and hydraulic presses. *Hydrokinetics,* the study of liquids in motion, is concerned with such matters as friction and turbulence generated in pipes by flowing liquids, and the use of hydraulic pressure in machinery.

Hydration. A reaction in which water is added to another substance, as in the conversion of an anhydride to its acid or the formation of a hydrate. The hydration of ions in solution is an example of solvation.

Hydrazine. A colourless oily toxic explosive liquid, H_2NNH_2. Hydrazine is a strong reducing agent used in rocket fuels, antioxidants, and corrosion inhibitors. M.pt. 2°C; b.pt. 113°C; r.d. 1.0 (25°C).

Hydride. A compound formemd between an element and hydrogen. Ionic hydrides are formed by electropositive metals and contain the negative *hydride ion* (H^-). An example is lithium hydride (LiH). Interstitial hydrides are the compounds formed by certain transition metals, such as palladium, when they absorb hydrogen. They are generally nonstoichiometric compounds. Nonmetals form covalently bonded compounds with hydrogen.

Hydriodic acid. *See* hydrogen iodide.

Hydrobromic acid. *See* hydrogen bromide.

Hydrocarbon. Any organic compound composed only of carbon and hydrogen. The hydrocarbons include alkanes, alkenes, alkynes, terpenes, and aromatic hydrocarbons, such as benzene and naphthalene.

Hydrocarbon. Any organic compound composed solely of Carbon and Hydrogen. Hydrocarbons include aliphatic compounds, in which the carbon atoms form a chain, and Aromatic Compounds, in which the carbon atoms form stable rings. The aliphatic group is divided into alkanes (e.g., Methane and Propane), alkenes, and alkynes (e.g., Acetylene), depending on whether the molecules of the compounds contain, respectively, only single bonds, one or more carbon-carbon double bonds, or one or more carbon-carbon triple bonds. Petroleum distillation yields useful fractions that are hydrocarbon mixtures, e.g., Natural Gas, Gasoline, Kerosene, home heating oil, lubricating oils, Paraffin, and Asphalt. Coal Tar is also a source of hydrocarbons. Hydrocarbon dervatives contain additional elements, e.g., oxygen,and include Alcohols, aldehydes, ketones, carboxylic acids, and halocarbons.

Hydrochloric acid. Chemical compound formed by dissolving hydrogen chloride (HCl) in water. Most hydrochloric acid produced has a concentration of 30% to 35% hydrogen chloride by weight. Hydrochloric acid is a strong acid and reacts with most common metals, releasing hydrogen and forming the metal chloride. The major use of hydrochloric acid is in the manufacture of other chemicals. It is also used is pickling (cleaning) metal surfaces, e.g., iron, before galvanizing.

Hydrochloride. A slat formed by reaction of an organic base with hydrochloric acid. Aniline, for example, forms aniline hydrochloride, $C_6H_5NH_2.HC1$.

Hydrocyanic acid. *See* hydrogen cyanide.

Hydrofluoric acid. *See* hydrogen fluoride.

Hydrofoil. Finlike device, attached by struts to the hull of a watercraft, that lifts the moving craft above the water's surface. The term also designates the vessel itself. Like an air-craft wing, the foil develops lift as it passes through the water; the hull is raised above the surface, and the reduced water drag permits greater speed. Some hydrofoil vessels are capable of traveling faster than 70 mph (113 km/hr). Hydrofoil vessels are used as ferries in Europe and Asia. A type of hydrofoil called a stabilizer is used on oceangoing passenger ships to minimize the effect of wave action on the vessel.

Hydrogel. A gel in which the dispersion medium is water.

Hydrogen. Symbol: H. A colourless odouless flammable gaseous element. It is the lightest element and also the most abundant in the universe, occuring in water and in most organic compounds. The elements is manufactured by several methods, including the Bosch process, the electrolysis of water, and the reaction of natural gas with steam. It is used in the Haber process for fixing nitrogen, the hydrogenatiuon of organic compounds, the reduction of some metal oxide ores, and in oxy-hydrogen welding. Chemically, hydrogen reacts directly with many other elements and all elements have hydrides. It is diatomic, H_2, and has ortho- and para- forms (*see* othohydrogen). There are three isotopes, the heavier ones being deuterium and tritium. A.N. 1; A.W. 1.007 97; m.pt. -259°C; b.pt. -252°C; r.d. 0.069(air = 1).

Hydrogen (H). Gaseous element, discovered by Henry Cavendish in 1766. The first element on the Periodic Table, Hydrogen is colorless, odorless, tasteless, slightly soluble in water, and highly explosive. The hot flame produced by a mixture of oxygen and hydrogen is used in welding, and in melting quartz and glass. Normal hydrogen is diatomic. The most abundant element in the universe, hydrogen is the major fuel in fusion reactions of the Sun and other Stars. Atmospheric hydrogen has three isotopes: *protium* (nucleus: one proton), the most common; *deuterium,* or heavy hydrogen (nucleus: one proton and one neutron), used in particle accelerators and as a tracer for studying chemical-reaction mechanisms; and *tritium* (nucleus: one proton and two neutrons), a radioactive gas used in the hydrogen bomb, in luminous paints, and as a tracer. Hydrogen's principal use is in the synthesis of Ammonia; liquid hydrogen has been greatly used as a rocket fuel, in conjunction with oxygen or fluorine. Deuterium oxide, or heavy water, is used as a moderator in nuclear reactors.

Hydrogenation. The reaction of a chemical compound (usually organic) with hydrogen. Common hydrogenations involve the additon of hydrogen to unsaturated molecules using catalysts. An example is the hydrogenation of ethylene: $C_2H_4 + H_2 = C_2H_6$.

Hydrogen bomb. Weapon deriving a large protion of its inergy from the nuclear fusion of hydrogen isotopes. In fusion, lighter elements are joined together to form heavier elements, and the end product weighs less than the components forming it. The difference in mass is converted into energy. Because extremely high temperatures are required to initiate fusion reactions, a hydrogen bomb is also known as a thermonuclear bomb. The presumable structure of a hydrogen bomb is as follow: an Atomic Bomb is surrounded by a layer of lithium deuteride (a compound of lithium and deuterium) and then by a tamper, or thick outer layer, frequently of fissionable material, that holds the contents together in order to obtain a larger explosion. The atomic explosition produces neutrons that fission the lithium into helium, tritium, and energy, and also produces the extremely high temperature needed for subsequent fusion of deuterium with tritium, and tritium with tritium. Explosion of the neutron bomb, which has a minimal atomic trigger and a nonfissionable tamper, produces blast effects and a hail or lethal neutrons but almost no radioactive fallout. The first thermonuclear bomb was exploded in 1952 at Enewetak by the U.S., the second in 1953 by the USSR.

Hydrogen bond. A weak attraction between an electronegative atom, such as oxygen, nitorgen, or fluorine, and a hydrogen atom that is covalently linked to another electronegative atom. Hydrogen bonds are electrostatic in origin: the electrons in the covalent bond are displaced towards the electronegative atom and electrostatic attraction occurs between the other electronegative atom and the unhielded positive charge of the hydrogen nucleus. Such bonds can be intermolecular: they occur between molecules of water, hydrogen fluoride, etc. They can also be intramolecular: hydrogen bonding is responsible for the structure of proteins and nucleic acids.

Hydrogen bromide. A colourless gas, HBr, made by passing hydrogen and bromine vapour over a platinum catalyst. It is a strong acid in aqueous solution (*hydrobromic acid*).

Hydrogen chloride. A colourless pungent gas HCI, prepared by reaction between concentrated sulphuric acid and sodium chloride. It is very soluble in water, forming strongly acidic solutions (*hydrochloric acid*).

Hydrogen cyanide (hydrocyanic acid prussic acid). An intensely poisonous colourless liquid, HNC, with a faint odour of bitter almonds, made by catalytic reaction between ammonia, air, and methane. It forms weekly acidic solutions in water. Hydrogen cyanide is used in making acrylates.

Hydrogen electrode. A half cell in which a platinum electrode is immersed in dilute acid of known concentration and hydrogen gas is bubbled through the solution over the platinum. The electrode reaction is $H^+ + e = H$: the hydrogen electrode is used as a standard for measuring electrode potentials.

Hydrogen fluoride. A colourless highly corrosive liquid, HF, used as a fluorinating agent and catalyst. It is prepared by treating calcium fluoride with sulphuric acid. In solution it forms *hydrofluoric acid,* which like hydrogen fluoride, attacks glass.

Hydrogen iodide. A colourless gas, HI mad by passing hydrogen and iodine vapour over a platinum catalyst. It is a reducing agent and a strong acid in aqueous solution (*hydriodic acid*).

Hydrogen peroxide. A colourless heavy liquid, H_2O_2, which is unstable and usually obtained in aqueous solution. It is a good oxidizing agent and is used as a rocket propellant, antiseptic, and bleach.

Hydrogen sulphide (sulphuretted hydrogen). A colourless flammable poisonous gas, H_2S, with an offensive odour of rotten eggs. It is a weak acid, made by addition of an acid to iron sulphide and used in metallurgy and analytical chemistry.

Hydrology. Study of water and its properties, including its distribution and movement in and through the land areas of the earth. The hydrologic cycle consists of the passage of water from the oceans into the atmosphere; onto, through and under the lands; and back to the ocean. Hydrology is mainly concerned with the part of the cycle that follows the precipitation of water onto the land and precedes its return to the oceans.

Hydrolysis. Chemical reaction of a compound with Water usually resulting in the formation of one or more new compounds. The most common hydrolysis occurs when a salt of a weak acid or weak base (or both) is dissolved in water. Water ionizes into negative hydroxyl ions (OH^-) and positive hydrogen ions (H^+), which become hydrated to form positive hydronium ions (H_3O^+). The salt also breaks up into positive and negative ions, and the formed ions recombine.

Hydrolysis. The reaction of a chemical compound with water. Usually the water is the solvent in which the reaction occurs. A common example is the hydrolysis of an ester to an alcohol and a carboxylic acid.

Hydrometer. An instrument for determining the relative density of a liquid typically consisting of a weighted bulb with a long graduated stem. The device floats vertically in the liquid and the length of stem immersed is proportional to its density.

Hydrometer. Calibrated glass float used to determine the Specific Gravity of a liquid. It usually consists of a thin glass tube that is weighted at one end so that it will float upright in a liquid. The scale reading on the tube that is level with the surface of the liquid in which the hydrometer floats indicates the number of time heavier or lighter the liquid is than water, i.e., the specific gravity of the liquid. The hydrometer is based on Archimedes' Principle.

Hydrophilic. Having an affinity for water.

Hydrophone. Device, used in Sonar appartus and in certain underwater weapons, that receives, underwater sound waves and converts them to electrical energy; the voltage generated can then be read on a meter or played through a loudspeaker. It is the marine equivalent of the Microphone.

Hydroponics. Growing of plants without soil, in water (and sometimes a sterile medium, e.g., sand) containing balanced concentrations of essential nutrients. Hydroponics can increase yields of commercial crops because plants are grown closer together than in the field; it also almost eliminates weeds and pests. The technique is limited by the support required to hold a given plant upright.

Hydroquinone (quinol). A white crystalline solid, $C_6H_4(OH)_2$, used as a photographic developer, antioxidant, and raw material for the manufacture of dyes.

Hydrostatics. The branch of mechanics concerned with the equilibrium of fluids.

Hydroxide. An ionic base containing the ion OH^- (the *hydroxide ion*).

Hydroxide. Chemical compound that contains the hydroxyl (–OH) radical. The term refers especially to inorganic compounds. Organic compounds that have the hydroxyl radical as a functional group are referred to as Alcohols. Most metal hydroxides are bases. Alkali Metal hydroxides, such as sodium hydroxide (NaOH), are strong bases and are very soluble in water. The Alkaline-Earth Metal hydroxides are less basic, and magnesium hydroxide (Milk of Magnesia) is only slightly basic. Some hydroxides, such as aluminum hydroxide [$Al(OH)_3$], exhibit Amphoterism.

Hydroxylamine. An explosive white crystalline solid, NH_2OH, made by reducing ammonium chloride and decomposing the hydroxylamine hydrochloride formed with a base. It is a reducing agent.

Hydroxyl group. The group -OH, as present in alcohols, phenols, and oxy acids.

Hyena. Chiefly nocturnal Mammal of the family Hyaenidae of Africa and SW Asia. Known for its cry, which sounds like maniacal laughter, it feeds mostly on carrion and can crush bones with its strong teeth and jaws. Three species are generally recognized: the spotted hyena (*Crocuta crocuta*) of sub-Saharan Africa; the smaller striped hyena (*Hyaena hyaena*) of Asia and N. Africa; and the brown hyena (*H. brunnea*) of S. Africa.

Hygrometer. Instrument used to measure the moisture content of a gas, as in determining the relative Humidity of the air. The most common type of hygrometer is the dry- and wet-bulb psychrometer. It consists of two identical mercury Thermometers, one of which has a wet wick around its bulb. The sling type of psychrometer is swung around in the air. Water evaporating from the wick absorbs heat from the bulb, causing the themometer reading to drop. The observer, after reading the dry-bulb temperature and the drop in the wet-bulb temperature, can determine the relative humidity from appropriate tables. Among other kinds of hygrometers are ones that use human hair or electrical resistance, rather than themometers, to determine moisture content.

Hygrometer. Any of various instruments for measuring humidity. Examples are the wet- and dry-bulb hygrometer, the dew-point hygrometer, and the hair hygrometer.

Hygroscopic. Denoting a substance that absorbs water from the atmosphere. Some solids can absorb enough moisture to form a solution (deliquescence).

Hyperbola. A conic with an eccentricity greater than 1. A. hyperbola is the locus of a point that moves so that the difference between its distance from two fixed points (the *foci*) is constant. In Cartesian coordinates its equation is $x^2/a^2 - y^2/b^2 = 1$, where a and b are constant and the curves are symmetric about the axes. It approaches the asymptotes $y = bx/a$ and $y = -bx/a$.

Hyperbolic. Having the shape of a hyperbola or a hyperboloid.

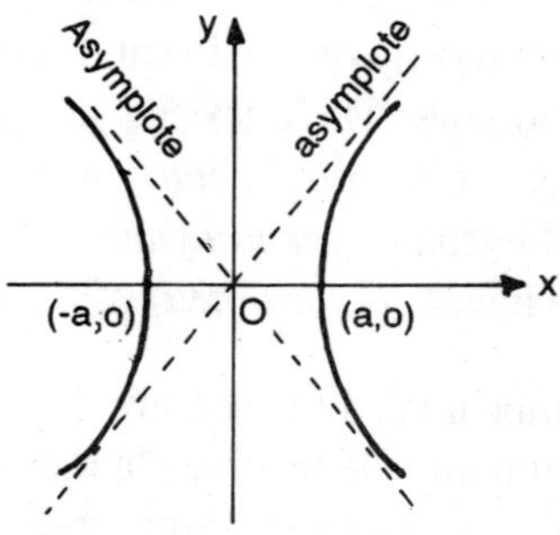

Fig. H-6. Hyperbola.

Hyperbolic function. One of a number of functions analogous to the trigonometric functions. The hyperbolic sine (written sinh x) is defined as $(e^x - e^{-x})/2$ and the hyperbolic cosine (cosh x) is $(e^x + e^{-x})/2$. The hyperbolic tangent (tanh x) is the ratio of sinh x to cosh x and the other three hyperbolic functions—cosech x, sech x, and coth x—are defined as reciprocals in the same way as the trigonometric functions.

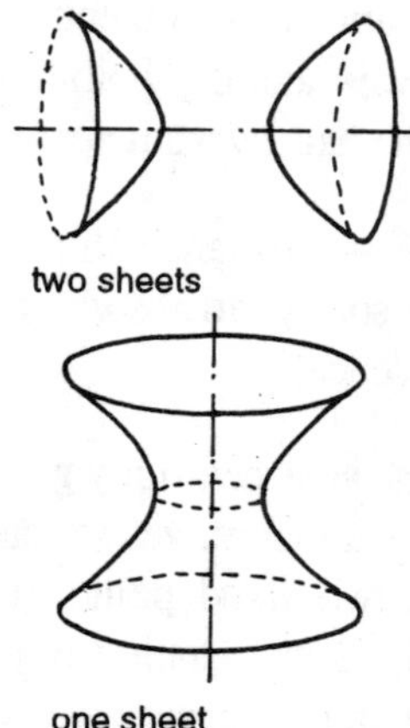

Fig. H-7. Hyperboloids.

Hyperboloid. Either of two types of surface that have hyperbolic cross sections in planes parallel to two of the coordinate planes and elliptical (or circular) cross sections in planes parallel to the third coordinate plane. A *hyperboloid of one sheet* ($x^2/a^2 + y^2/b^2 - x^2/c^2 = 1$) is a single surface with its elliptical cross sections in the *xy* plane. A *hyperboloid of two sheets* is made up of two surfaces and the ellipitical cross sections are in the *yz* plane. Simple forms of these hyperboloids, with circular cross sections, can be generated by rotating a standard hyperbola about its *y* or *z* axis respectively.

Hypercharge. Symbol: *Y*. A property of elementary particles described by a quantum number equal to the sum of the particle's baryon number and its strangeness. Hypercharge is conserved in strong and electromagnetic interactions: not in weak interactions.

Hyperconjugation. Electronic interaction occurring between a double bond in a compound and carbon-hydrogen single bond in a carbon-hydrogen single bond in an attached alkyl group. Under certain circumstances the double bond and the single C-H bond appear to behave rather like conjugated double bonds.

Hyperfine structure. *See* fine structure.

Hypermetropia. A defect of the eye in which parallel rays are focused to a point behind the retina when the eye is at rest so that distant objects are indistinct and the eye has to accommodate to bring them to focus. In hypermetropia the distance between the front and back of the eyeball is too short and the condition can be corrected by convex spectacle lenses. Hypermetropia is sometimes called *long sight,* which should not be confused with far sight (presbyopia).

Hyperon. Any of a lcass of elementary particles that have half-integral spin, take part in strong interactions, and have masses larger than those of the proton and the neutron. The hypersons are thus all baryons that are not nucleons. They include the lambda, xi, sigma, and omega-minus particles. All decay into nucleons.

Hypersonic. Relating to a speed in excess of Mach 5.

Hypertension. Elevated blood pressure resulting from an increase in the amount of blood pumped by the heart or from increased resistance to the flow of blood through the small arterial blood vessels (arterioles). When the cause is unknown, the condition is called primary, or essential, hypertension. When a cause can be identified (e.g., a disorder of the adrenal glands, kidneys, or arteries), the condition is known as secondary hypertension. Factors such as heredity, obesity, high salt intake, and emotional stress are thought to play a role. Known as the "silent killer," hypertension often produces few overt symptoms; it may, however, result in damage to the heart, eyes, kidneys, or brain and ultimately lead to heart failure or Stroke. Treatment of hypertension includes weight- and salt-control diets and various drugs, such as Diuretics and Beta Blockers.

Hypnosis. Term for a psychological state superficially resembling sleep and generally induced by the monotonous repetition of words and gestures while the subject is completely relaxed. Most people can be hypnotized, although individuals vary greatly in susceptibility to hypnosis. The hypnotic state is characterized by heightened suggestibility e.g., the hypnotized individual can be induced to follow many (but not all) instructions or to endure painful conditions. Hypnosis is occasionally used as an aid in medical practice and psychotherapy (e.g., in helping Amnesia patients recall traumatic events) and in behavior therapy (see Behavior Modification) to assist people in breaking unwanted habits (e.g., smoking, overeating).

Hypochlorite. Any salt or ester or hypochlorous acid.

Hypochlorous acid. A greenish-yellow liquid, HOC1, known only in aqueous solution and in the form of hypochlorite salts. It is an oxidizing agent formed by the action of chlorine on water.

Hypocycloid. A curve that is the locus of a point on the circumference of a circle as that circle rolls on the inside of another circle. *Compare* epicycloid.

Hypoglycemia. Abnormally low level of blood glucose (sugar), the body's chief energy source (hence the term *low blood sugar*). It is most often caused by an oversecretion of insulin from the pancreas triggered by stress, exercize, fasting, or disorders of the adrenal or pituitary glands, liver, or pancrease; in persons with diabetes it may result from an overdose of insulin. Symptoms range from weakness, fatigue, shakiness, and anxiety to mental disturbances, convulsions, and coma. Acute episodes are relieved by ingestion of glucose or sucrose. Long-term treatment involves control of causative factors and diet regulation.

Hypophosphite. *See* phosphite.

Hypophosphorous acid. *See* phosphorous acid.

Hypothalamus. Important supervisory center in the Brain. The hypothalamus regulates body temperature, Blood Pressure, heartbeat, Metabolism of fats and carbohydrates, and blood sugar level. It participates, with other brain centers, in the control and monitoring of several other functions, such as water balance and certain motor responses.

Hypothesis. A proposition explaining observed facts. A hypothesis may be a highly probably explanation of the facts. Alternatively it may be a *working hypothesis*, entertained as a guide to further experiment and investigation.

Hysterectomy. Surgical removal of the uterus, sometimes including the removal of the cervix, Fallopian tubes, and ovaries (see Reproductive System). It is performed in cases of malignant tumors, or benign growths causing bleeding and pain. Hysterectomy does not interfere with sexual activity, but since it eliminates the possibility of childbearing, it is avoided in younger women when more conservative treatment is feasible.

Hysteresis. An apparent lag of an effect with respect to the magnitude of the agency producing the effect. *Magnetic hysteresis* is the lag of the magnetization of a ferromagnetic material with respect to the external field producing it. It can be displayed by plotting a graph of the magnetic flux density *B* inside the sample produced by an external magnetic field strength *H*. If the sample is initially unmagnetized, the flux density increases steadily up to a point (A) at which the flux density is increasing linearly with field strength. If the field strength is then decreased to zero, the curve *OA* is not retraced. Instead, the flux density decreases less markedly and when the external magnetizing field is zero, it has a residual value given by the intercept *OB*. This value of *B* is the *remanence* of the material. As the value of *H* is increased in the oppoiste direction, the value of *B* falls further until at *C* it reaches zero again. The external field necessary to remove the magnetization is the material's *coercivity*. If *H* is increased still further in this direction a second saturation (D) is reached. The field is again reduced to zero, reversed, and increased in the opposite direction, giving the complete curve ABDCEF. This is the *hysteresis loop* of the material. A graph of magnetization (*M*) against external field has the same form but differs in that at saturation *M* becomes constant with increasing *H*.

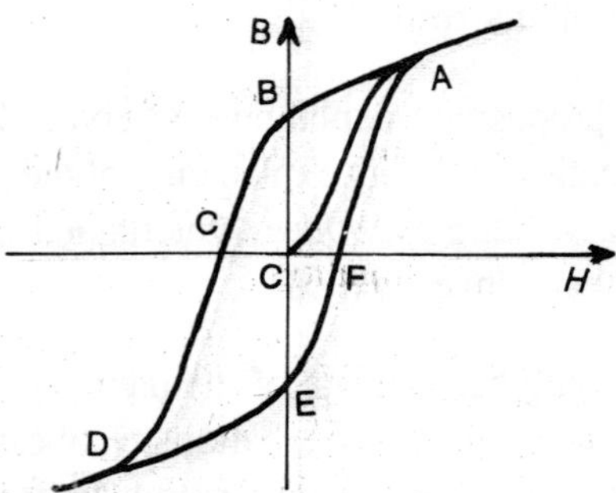

Fig. H-8. Hysteresis loop.

The magnetization and demagnetization of a sample according to such a cycle requires energy and it can be shown that the energy and it can be shown that the energy necessary is equal to $\mu_0 A$, where *A* is the area of the hysteresis loop and μ_0 the magnetic constant. The process causes a dissipation of energy, known as *hysteresis loss*, in systems in which a ferromagnetic material is being magnetized and demagnetized by a varying field.

Magnetic materials used in generators, transformer cores, and electromagnets require low coercivity and low hysteresis loss (i.e.their hysteresis loops should have a small area). Permanent magnets require materals with high remanence and coercivity.

Other types of hysteresis are known. *Electric hysteresis* for example is an effect shown by materials exhibiting ferroelectricity in which a graph of displacement against electric field strength shows a similar form.

Hysteria. A neurotic condition in which a psychological conflict is converted into a bodily disturbance. Hysterical symptoms, which have no underlying physical cause, may include limb paralysis, blindness, convulsive seizures, or, when expressed as a behavioural rather than physical infirmity, Amnesia. Pioneering work on the psychological basis of hysteria was done by Josef Breur and Sigmund Freud.

I

Iceberg. Mass of ice that has become detached from an ice sheet or Glacier and is floating in the ocean. Only about one ninth of its total mass projects above the water. Rocks dropped to the ocean floor by melting icebergs have been studied to determine the range of icebergs during glacial periods. Because of the 1912 collision of the ocean liner Titanic with an iceberg, a constant census of icebergs is maintained and their locations reported to nearby ships.

Icebreaker. Ship of special hull design and wide beam, with a relatively flat bottom, designed to force its way through ice. When the icebreaker charges into the ice at full speed, its sharply inclined bow, meeting the edge of the ice, slides onto it, and the weight of the vessel causes the ice to collapse. Able to force their way through ice up to 35 ft (10.7 m) thick, icebreakers have been important in keeping northern waterways open.

Ice Cream. Sweet frozen food, churned from milk fat and solids, sugar, flavoring, a stabilizer (usually gelatin), and sometimes eggs, fruits, or nuts. Marco Polo brought back reports of flavored, iced foods from the Fat East. From Italy the confection spread to France and England, reaching America early in the 18th cent. Commercial manufacture of ice cream began in the U.S. in 1851 and has become an important industry.

Iceland Spar. Colorless variety of crystallized Calcite, known for its properties of transparency and double refraction, found primarily in Iceland. It is used chiefly in the manufacture of Nicol prisms.

Ice Point. The temperature at which ice melts. It is taken to be the temperature at which ice and water are in equilibrium at standard pressure and is used as a fixed point on 'temperature scales.

Icosahedron. A polyhedron with twenty faces. A *regular icosahedron* has congruent equilateral triangles as its faces.

Ideal Gas (perfect gas). An idealized gas composed of atoms that have a negligible volume and undergo perfectly elastic collisions with one another and with the walls of their container. Such a gas would obey the gas laws under all conditions. Real gases behave like ideal gases at low pressures.

Identity. An equation that holds for all values of the variables in the expression. For instance,

$$(x-y)^2 = x^2 + y^2 - 2xy$$

is an identity. $x^2 + y^2 = y^2$ is not. It is only true when $x = 0$.

Ignis Fatuus. Will-o'-the wisp.

Iguana. Large Lizard (family Iguanidae) found in tropical America and the Galapagos Islands. The common iguana (*iguana iguana*), a tree-living species found along streams from Mexico to N South America, is bright green with a crest of spines from the neck to the striped tail. Its tail accounts for two thirds of its length (3-6 ft/90-180 cm).

Ilmenite. A black mineral consisting of a mixed iron and titanium oxide, $FeO.-TiO_2$, used as an ore of titanium.

Image. The visual impression of an object produced by a lens, mirror, or optical system. A *real image* is formed by convergence of rays of light and can be produced on a screen. A *virtual image* is one from which the rays of light appear to diverge. Such an image could not be projected onto a screen.

Image Converter. An electronic device for obtaining a visible image from an image formed with infrared or other nonvisible radiation. The radiation is focused onto a photocathode and the electrons released are accelerated and focused onto a fluorescent screen.

Image Intensifier. An apparatus for intensifying the brightness of an optical image, consisting essentially of a form of photomultiplier in which the optical image releases photoelectrons which are multiplied and directed onto a fluorescent screen.

Image Orthicon. A type of television camera tube in which the image is produced on a photocathode. Each part of this electrode emits electrons in proportion to the amount of incident light and these in turn hit a target screen, causing secondary emission. The process creates an electrostatic image on the target in which the amount of positive charge at each point corresponds to an amount of incident light. The back of the target is scanned by an electron beam and the reflected beam detected. Its intensity depends on the charge on the target at each point.

Imaginary. Denoting a number that is the square root of a negative number. Imaginary numbers are expressed in the form ai. where a is a real number and i is the square root of minus one.

Imide. An organic compound containing the group -CO.NH.CO- (the *imido group*).

Imido Group. Imide.

Imine. An organic compound containing the group. -NH-, where the nitrogen is not linked to carbonyl groups or hydrogen atoms. The group is the *imino group*.

Imino Group. Imine.

Immersion Objective. A microscope objective used in an oil-immersion microscope.

Immiscible. Denoting liquids that do not mix completely to form a homogeneous mixture. Oil and water, for example, are immiscible—neither component will dissolve in the other.

Immunity. Ability of an organism to resist disease by production of antibodies. The introduction into the body of a foreign substance called an *antigen*—a virus or bacterium or some nonliving substance—triggers the production of antibodies by special white blood cells called lymphocytes, located chiefly in the spleen and lymph nodes. The process is highly specific, with different lymphocytes recognizing only certain antigens and producing antibodies against only that particular antigen. The antibodies act to neutralize the effect of the antigen by removing the antigen from the body's circulation, by coating the antigen or antigen-producing agent so that it can be more easily destroyed by the body's scavenger cells, or by combining with the antigen in such a way as to make it incapable of infecting body cells. The body also resists invading agents by phagocytosis, a process by which special white cells (macrophages and others) engulf and destroy the invading microorganism. Immunity has been found to be important not only in resisting infectious disease, but also in the defense against Cancer, in successful organ Transplantation, in Allergies, and in Autoimmune Diseases (the result of an organism producing antibodies against its own proteins and cell components). Immunosuppressive Drugs (substances used to prevent the production of antibodies) are used to treat some autoimmune diseases and to prevent the body's rejection of organ transplants.

Immunology. Study of the resistance of organisms to infection. Immunologists study behavior of pathogenic (disease-causing) organisms, factors that enable the host to resist infection, and defensive measures used by organisms to fight invading pathogens.

Immunosuppressive Drug. Substance that suppresses the activity of lymphocytes, the white blood cells that form antibodies (the body's response to foreign substances. Such drugs, including Cortisone and other Steroids, are administered to prevent rejection by a recipient's body of an organ transplanted from a donor. Other immunosuppressives act by interfering with Nucleic Acid synthesis. These drugs, which are especially effective against rapidly proliferating cells, are used in the treatment of cancer particularly leukemia.

Impedance. In electricity, measure of the degree to which an electric circuit resists electric-current flow when a voltage is impressed across its terminal. Impendance, expressed in Ohms, is the ratio of the voltage impressed across a pair of terminals to the current flow between those terminals. In direct-current (DC) circuits, impedance corresponds to Resistance. In alternating-current (AC) circuits, impedance is a function of resistance, Inductance, and Capacitance. Inductors and Capacitors build up voltages that oppose the flow of current. This opposition, called reactance, must be combined with resistance to find the impedance. The reactance produced by inductance is proportional to the frequency of the alternating current, whereas the reactance produced by capacitance is inversely proportional to the frequency. In order to transfer maximum electrical power from one device to another, the two impedances must be matched.

Impedance. Symbol : Z. A measure of the ability of an alternating-current circuit to resist the flow of current, equal to the potential diffence across the circuit divided by the current through it. The voltage and the current will both be varying quantities and, in general, will not be in phase. The impedance is a complex quantity: $Z = R + iX$, where R is the resistance and X is the reactance. It can also be put in the form $Z = (V_0/I_0)_e$, where V_0 and I_0 are the amplitudes of the voltage and the current and θ is the phase angle between them. The modulus of the impedance. $|Z|$, is given by $(R^2 + X^2)^{1/2}$. Impedance is measured in ohms.

Imperial Units. The system of weights and measures used in Britain, based on the pound and the yard.

Impetigo. Highly contagious skin infection affecting mainly infants and children, caused either by hemolytic streptococcus or staphylococcus. The rash consists of red spots or blisters that rupture, discharge, and become encrusted. Antibiotic ointment is usually effective.

Implosion. Violent collapse inwards of an evacuated container, caused by the external pressure.

Impotence. Inability of the male to perform sexual intercourse. Impotence can result from psychological factors or such physical causes as hormonal abnormalities, Diabetes, Drugs, and Alcoholism. Treatment depends on the under-lying cause. Impotence should be distinguished from Sterility (inability to produce sperm adequate for reproduction).

Improper Fraction. Fraction.

Impulse. A force acting for a very short time, as in an impact. Its magnitude is the product of the force and the time during which it acts. An impulse on a body is equal to the change of momentum that it produces.

Incandescence. The emission of visible light from a hot object.

Incense-tree. Common name for deciduous shrubs and trees of the family Burseraceae, found chiefly in tropical America and NE Africa. The incenses frankincense and myrrh are prepared from the resin exuded by some species. Frankincense, or olibanum, derived mainly from *Boswellia carterii,* is used medicinally and for fumigation. Myrrh, prepared especially from *Commiphora erythraea* and common myrrh, *C. myrrha,* is used medicinally and in perfumes, and was used by the ancients in embalming. Frankincense and myrrh, together with Gold, were the gifts of the Wise men of the East.

Inch. Abbrev.: in. An Imperial unit of length equal to one twelfth of a foot. It is equivalent to 0.0254 metre.

Incidence. Impingment of a beam of light or other radiation on a surface. The *angle of incidence* is the angle between the beam and the normal to the surface at the point of incidence.

Inclination. Dip.

Inclinometer. Dip Circle.

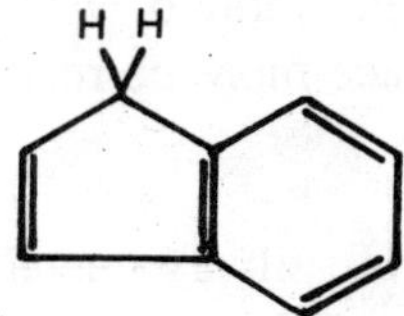

Fig. I-1. Indene.

Indene. A colourless flammable liquid heterocyclic compoumd, C_9H_8, obtained from coal tar and used in the manufacture of synthetic resins. M.pt. -35°C; b.pt. 182°C; r.d. 1.01.

Independent Variable. Variable.

Index. A number indicating the power to which a number or expression is raised. In x^7, 7 is the index.

Indicator. A substance used to indicate the presence of another compound by a characteristic colour change. Starch, for example, indicates iodine by forming a blue complex. The term is usually used for compounds that are suitable for showing the end point of a titration by having one colour in the presence of excess of one reactant and another colour in excess of the other and sharp colour change between the two. Indicators employed in acid-base titrations are often dyes that have one molecular structure in excess H^+ ions and another in excess OH^- ions. Examples are litmus, phenolphthalein, and methyl orange.

Indicators, Acid-base. Organic compounds that in water solution exhibit colour changes indicative of the acidity or basicity of the solution. Litmus, for example, is red in acidic solution and blue in basic. Other common indicators are phenolphthalein and methyl orange.

Indium (In). Metallic element, discovered spectroscopically in 1863 by Ferdinand Reich and H.T. Richter. Soft, malleable, ductile, lustrous, and silver-white, it remains liquid over a wide temperature range. Indium wets glass and can be used to form a mirror surface more corrosion-resistant than one of silver.

Indium. Symbol: In. A soft silvery metallic element present in small amounts in some zinc ores. Its is used in alloys for jewellery and dental fillings. A.N. 49; A.W. 114.82; m.pt. 156.4°C; b.pt. 1450°C; r.d. 7.31; valency 3.

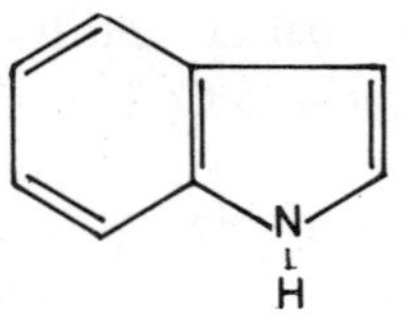

N
H

Fig. I-2 Indole.

Indole. A white to yellowish crystalline solid, C_8H_7N, obtained from coal tar. It is used in medicine and in perfumery. M.pt. 52°C; b.pt. 254°C.

Inductance. A measure of the ability of an electric circuit to resist the flow of a changing current as a result of an induced e.m.f. opposing the current flow.

Inductance. Quantity that measures the electromagnetic Induction of an Electric-Circuit component. The self-inductance L of a circuit component determines the magnitude of the Electromotive Force (emf) induced in it as a result of a given rate of change of current through the component. The mutual inductance M of two components, one in each of two separate but closely located circuits, determines the emf that each may induce in the other for a given current change. Inductance is expressed in units of henrys. A device designed to produce an inductance, *e.g.*, a wire coil, is called an inductor.

Induction. Any change in a body produced (induced) by the action of a field. Electrostatic induction, magnetic induction, electromagnetic induction.

Induction. In Electricity and Magnetism, common name for three distinct phenomena. *Electromagnetic induction* is the production of an Electromotive Force (emf) in a conductor as a result of a changing magnetic Field about the conductor. Such a variation may be produced by relative motion between the conductor and the source of the magnetic field, as in an electric Generator, or by varying the strength of the entire field. Changing the current in a given circuit can also induce emf in a nearby circuit unconnected with the original circuit; this is called mutual induction and is the basis of the Transformer. *Electrostatic induction* is the production of an unbalanced electric Charge on an uncharged metallic body as a result of a charged body being brought near it without touching it. If the charged body is, *e.g.*, positively charged, electrons in the uncharged body will be attracted toward it; if the opposite end of the body is then grounded, electrons will flow into it to replace those drawn to the other end. The body thus acquires a negative charge after the ground connection is broken. *Magnetic induction* is the production of a magnetic field in a piece of unmagnetized iron or other ferromagnetic substance when a magnet is brought near it. The magnet causes the individual particles of the iron, which act like tiny magnets, to line up so that the sample as a whole becomes magnetized.

Induction. The process of reasoning from the paticular to the general. Francis Bacon proposed inductive logic as the logic of scientific discovery, holding Deduction as the logic of argumentation. In fact, both processes are used together regularly in the empirical sciences: by the observation of particular events (induction) and from already known principles (deduction), new hypothetical principles are formulated and laws induced.

Induction coil. A device for producing a high pulsed voltage. It consists of a laminated iron core around which are wound two coaxial coils, a primary coil consisting of a few turns of wire and a secondary coil having a large number of turns. A low current is supplied to the primary, continuously interrupted by a mechanical vibrator, and a high pulsed e.m.f. is induced in the secondary coil.

Induction Heating. A method of heating metals by the eddy currents induced in them by an alternating magnetic field.

Induction Motor. Electric motor.

Inductive. Involving electromagnetic induction or inductance.

Inductor. A choke or other component used to introduce inductance into a circuit.

Inequality. A mathematical statement that one quantity is greater or less than another. Inequalities are written in the forms $a > b$ (a is greater than b) and $a < b$ (a is less than b).

Inert Gas or Noble Gas. Any of the elements in group 0 of the Periodic Table. In order of increasing atomic number, they are Helium, Neon, Argon, Krypton, Xenon, and Radon. Sometimes called the rare gases (although argon makes up 1% of the atmosphere), they are colorless, odorless, and tasteless. Inert gases have very low chemical activity because their outermost, or valence, electron is complete, containing two electrons in the case of helium and eight in the remaining cases.

Inert. Denoting a chemical substance that is very unreactive.

Inert Gas (rare gas, noble gas). Any of a class of elements belonging to group O of the periodic table. The inert gases have a filled shell of electrons and were once thought to form no compounds. Xenon and, to a lesser extent, krypton are now known to form compounds, especially with fluorine.

Inertia. In physics, the resistance of a body to any alteration in its state of Motion, *i.e.*, the resistance of a body at rest to being set in motion or of a body in motion to any change of speed or of direction of motion.

Inertia. The tendency of a body to resist changes in its motion.

Inferior Planet. Either of the two planets, Mercury and Venus, whose orbits around the sun lie within that of the earth. *Compare* superior planet.

Infinity. In mathematics, that which is not finite; it is often indicated by the symbol ∞. A Sequence of numbers is said to approach infinity if the numbers eventually become arbitrarily large, *i.e.,* larger than any specified number. The word *infinite* is also used to describe a Set with more than any finite number of elements, *e.g.,* the set of points on a line or the set of all prime numbers. Georg Cantor showed that there are different orders of infinity, the infinity of points on a line being of a greater order than that of prime numbers, and developed the theory of transfinite numbers as a means of distinguishing them.

Influenza. Acute, highly contagious disease caused by a number of different viruses. The disease usually begins abruptly with fever, muscular aches, and inflammation of the respiratory mucous membranes; its more severe forms are bacterial Pneumonia and Bronchitis. Influenza epidemics have decimated large populations; an outbreak in 1981 killed over 20 million people. An injection with influenza virus vaccine can confer temporary immunity.

Information Storage and Retrieval. The systematic process of collecting and cataloging data so that it can be located and displayed upon request. Computers and Data Processing have made possible the high-speed, selective retrieval of large amounts of information in such diverse fields as banking, law enforcement, jurisprudence, and medicine. There are several basic types of information-storage-and-retrieval systems. *Document-retrieval systems* store entire documents; these are usually retrieved by title or through a series of key words that denote the category to which they belong. Full text searching, the capability of retrieving on the basis of any word in the document, is becoming increasingly common, particularly in legal research. *Data-base systems* store the information as a series of discrete records that are, in turn, divided into discrete fields (*e.g.*, name, height, and weight); records can be searched and retrieved on the basis of the content of the fields *e.g.* all people who weigh more than 150 lb/68 kg). The data is stored within the computer, either in main storage or auxiliary storage, for ready access. Upon retrieval it may be delivered on magnetic tape or microfilm, or it may be printed. *Reference-retrieval systems* store references to documents; references relevant to a particular request are retrieved and printed in a list. Such a system can therefore provide an indix to literature from a wide variety of sources (*e.g.*, books, periodicals, technical journals) on a particular subject (*e.g.*, obesity).

Information Theory. Mathematical theory that explains aspects and problems of information and communication. In this theory, the term *information* is a measure of the freedom of choice with which a message is selected from the set of all possible messages. Information is distinct from meaning, because a string of nonsense words and a meaningful sentence may be equivalent with respect to information content. Numerically, information is measured in bits (short for *binary digits*). One bit is equivalent to the choice between two equally likely choices. When several choices are equally likely, the number of bits is equal to the Logarithm of the number of choices taken to the base two. When the various choices are not equally probable, the situation is more complex. The mathematical expression for information content closely resembles the expression for Entropy in thermodynamics. The greater the information in a message, the lower its randomness, or "noisiness," and hence the smaller its entropy. A message proceeds along some channel from the source to the receiver; information theory defines for any given channel a limiting capacity or rate at which it can carry information, expressed in bits per second. The theory succeeds remarkably in outlining the engineering requirements and limitations of communications systems.

Infrared Radiation. Electromagnetic Radiation having a wavelength in the range of 750 to 1,000,000 nanometers, thus occupying that part of the electromagnetic spectrum with a frequency less than that of red visible Light and greater than that of Microwaves. Infrared radiation is thermal, or heat, radiation, and is produced by any body having a temperature above absolute zero. It has many of the same properties as visible light, such as being reflected or refracted.

Infrared Radiation. Electromagnetic radiation with wavelengths in the range 730 nm to about 1 mm. Infrared radiation lies between visible radiation and radio waves (or microwaves) in the electromagnetic spectrum. It is roughly classified into the *near infrared* (730 nm-75 μm) and the *far infrared* (>75 μm). Infrared radiation is emitted by hot bodies: the emission involves changes in vibrational energy of atoms and molecules. When absorbed by bodies it causes an increase in the vibrational energy: *i.e.* it is converted directly into heat. It is sometimes called *heat radiation* or *radiant heat.*

Infrared Spectrum. An emission or absorption spectrum in the infrared region. Absorption of infrared radiation occurs by a change of a molecule from one vibrational energy level to another. The spectrum gives information on both the molecule's vibrational frequencies and its shape. Infrared spectroscopy is widely used in organic chemistry to identify the presence of particular groups of atoms by their frequency of vibration.

Infrasonic. Below audible frequencies: *i.e.* below about 20 Hz.

Infrasound. Propogated vibrations with a frequency below that of audible sound: *i.e.* below about 20 Hz.

Ink. Pigmented fluid used for writing or drawing, or viscous compound used for printing. The oldest known variety, India or China ink, is still used with brushes in the Orient. Plant dyes and other fluids have been used as ink. The black and blue-black inks used today date from as early as the 2d cent. Soluble dyes allow modern inks to be filtered through writing instruments. Newsprint ink is absorbed into the paper; many other inks evaporate, leaving pigment on the surface.

Inlaying. Process of ornamenting a surface by setting into it material of a different color or substance. Of ancient origin, inlaying of materials such as wood, stone, ivory, glass, metal, and mother-of-pearl has been used on a wide variety of architectural and decorative objects.

Inorganic. Relating to compounds that do not contain carbon: *i.e.* compounds that are not organic. The study of such compounds is the province of *inorganic chemistry.*

Input. 1. The signal, current, voltage, etc., fed into an electrical circuit or device.

2. The information fed into a computer. The term also refers to the part of the computer system that converts the information into the form in which it is stored. Input may be by punche cards or paper tape—the card or tape reader converts the information by a light beam and photocell arrangement which produces a series of current pulses. Magnetic tape can also be used as the input medium. In optical character recognition the information is in the form of conventional printed characters.

Insect. Invertebrate animal of the class Insecta, an Arthropod. There are more than 700,000 known species. Insects have a chitinous exoskeleton; a segmented body composed of a head (with three pairs of mouthparts, compound and simple eyes, and sensory antennae); a trisegmented thorax (with three pairs of jointed legs and two pairs of wings); and an abdomen (with reproductive appendages). They breathe through a complex network of air tubes (tracheae) opening on the sides of the body. Reproduction is usually sexual, although Parthenogenesis is also common. Some 80% of insect species undergo complete Metamorphosis. Insects are both harmful to mankind (as disease carriers and agricultural pests) and beneficial as pollinating agents, as predators on harmful species, and as food in some parts of the world. Fossil records show that insects have undergone relatively little change for 200 million years. The class includes the Beetles; the Moths and Butterflies; the Wasps, Ants, and Bees; the Flies and Mosquitoes; the true Bugs; Aphids; and the Cockroaches and Grasshoppers.

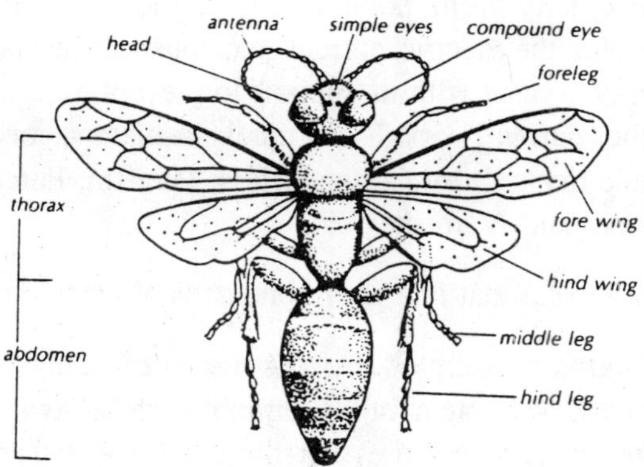

Fig. I-3 Insect: General anatomy of a wasp, an insect of the order Hymenoptera.

Insecticide. Chemical agent used to kill insect pests. Insecticides have helped increase the yield and improve the quality of crops, but there has been concern about the dangers of insecticide residues in the ecosystem and in foodstuffs. Such concerns have led to governmental regulation and the replacement of toxic insecticides that persist in the environment (*e.g.,* DDT and other chlorinated hydro-carbons) by compounds that break down more quickly into nontoxic forms. In addition, partly because insects can become resistant to insecticides, there has been interest in comprehensive pest-management techniques, including the use of biological control agents (*e.g.,* predator insects) that harm only certain pests.

Insoluble. Denoting a substance that does not dissolve. Unless the solvent is specified it is understood to be water.

Instantaneous. Denoting a value of a changing physical quantity at a given instant of time.

Insulation. Use of materials to inhibit or prevent the Conduction of heat or of electricity. Common heat insulators are Asbestos, Cellulose fibers, Feathers, Fiber Glass, Fur, Stone, Wood, and Wool; all are poor conductors of heat. In the conduction of electricity from point to point, the conductor acts as a guide for the electric current and must be insulated at every point of contact with its support to prevent escape, or leakage, of the current. Good electrical insulators, or Dielectrics, include dry air, dry Cotton, Glass, Paraffin, Porcelain, Resin, Rubber, and Varnishes.

Insulator. A solid that is a poor conductor of electricity or of heat.

Insulin. Hormone secreted by the islets of Langerhans in the Pancreas. A protein of 21 amino acids in two cross-linked chains, insulin was the first protein to be sequenced (1951) and its synthesis was reported by several groups in the mid-1960s. In general, insulin acts to reduce levels of glucose in the blood by interacting in some unknown way with cell membranes. It also increases protein synthesis in muscle. Insufficient insulin in the body results in Diabetes, a condition treated by the administration of insulin.

Intaglio. Design cut into stone or other material, or etched or engraved in a metal plate, producing a concave effect that is the reverse of relief or Cameo. Intaglio Printing techniques include Engraving and Etching.

Integral Calculus. The branch of calculus concerned with integration.

Integrated Circuit. An electronic circuit made in a single small unit. Typically, an integrated circuit may be constructed on a small slice of silicon (called a *chip*) by forming a network of junction or field-effect transistors on its surface. Interconnections between the elements are made by thin metal films.

Integrated Circuit. Electric Circuit or module packaged as a single unit with leads extending from it for input, output, and power-supply connections. All the electronic devices are formed by selective treatment (doping) of a single chip of Semiconductor material. Integrated circuits are categorized according to the number of Transistors or other active circuit devices they contain; an active circuit device is one that receives power from a source other than its input signal. An ordinary integrated circuit (IC) may contain up to several tens of such devices; a medium-scale integrated circuit (MSI), many tens to several hundred; a large-scale integrated circuit (LSI), several hundred to a few thousand; and an extra-large integrated circuit (ELSI), a few thousand or more.

Integration. A mathematical operation involving determination of the sum of a number of terms as each term becomes infinitesimally small and the number of terms increases to infinity. A common application is in finding the area under a curve. For instance the area between the curve $y = f(x)$, the x axis, and two ordinates $x = a$ and $x = b$, can be found by dividing it into small rectangles by a number of equally spaced ordinates. If these are separated by a distance Δx, each has an area $y\Delta x$. The limit of the sum of these areas, as Δx tends to zero, is the area under the curve, written $\int_m^n y dx$.

This is the *integral* of y with respect to x. Integration is the inverse operation of differentiation.

Intelligence. The general mental ability involved in processes such as calculating, reasoning, classifying, learning, the use of language, and adjusting to new situations. It is widely accepted that although potential intelligence is related to heredity, the environment is a critical factor in determining the extent of its expression. Since the concept of intelligence is elusive, it has generally been defined by psychologists as that which is measured by intellgence tests. Alfred Binet and Theodore Simon developed the first modern intelligence test in 1905 for the French school system. Currently the most widely used intelligence tests are the Wechsler Intelligence Scale for Children (WISC) and the Wechsler Adult Intelligence Scale (WAIS). Others commonly used in the past include the Army Alpha Test, first administered to recruits in World War I, and the Revised Standford-Binet Intelligence Tests, developed by Lewis Terman in the 1930s. In response to the criticism that intelligence is an attribute of the entire personality and thus cannot be measured in isolation, there has been a decline in interest in pure intelligence tests in favor of tests that also measure special aptitudes and personality factors.

Interaction. Electromagnetic interaction, strong interaction, weak interaction, gravitational interaction.

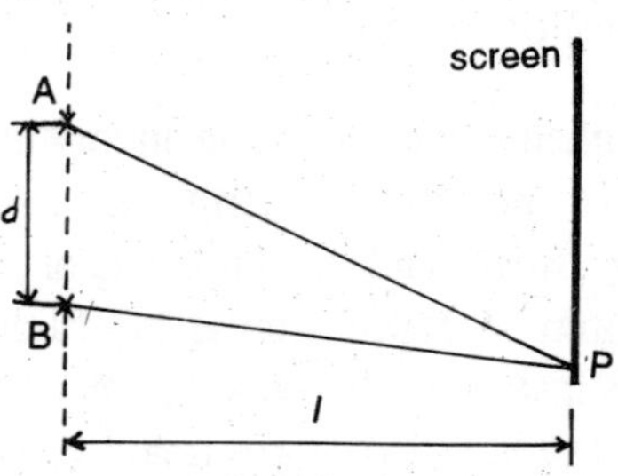

Fig. I-4. Interference.

Interference. In physics, the effect obtained when two systems of Waves reinforce, neutralize, or in other ways interfere with each other. Interference is observed in waves both in a material medium (such as Sound) and in Electromagnetic Radiation. *Constructive interference* occurs when two waves in the same phase combine. The waves reinforce each other, and the amplitude of the resulting wave is equal to the sum of the amplitudes of the interfering waves. When the phases of the two waves are shifted over 180°, *i.e.*, the maximum positive amplitude of one wave coincides with the maximum negative amplitude of the other waves, *destructive interference* occurs, which results in the canceling of the waves when they have the same amplitude.

Interference of Light. Interation of two sets of light waves with the same frequency and phase to produce light and dark bands (called *fringes*). If, in the illustration, A and B are monochromatic coherent sources of light with the same frequency and intensity, interference fringes can be observed on the screen a distance I away. A light wave can be thought of as a train of alternate maxima and minima. At the point P, if two maximum values or minimum values coincide constructive interference occurs and a bright fringe is formed. If a maximum coincides with a minimum, destructive interference results and a dark fringe is produced. For a bright fringe the path difference BP - AP must equal a whole number of wavelengths $n\lambda$, where *n* is an integer. The separation of two fringes is $\lambda I/d$, where *d* is the separation of the sources. Interference fringes of this type can be formed by using a single light source and a pair of slits or a biprism.

If the light is not monochromatic coloured fringes are formed. Interference effects of this type are seen in soap bubbles and thin films of oil. The constructive and destructive interference occurs between trains of light waves reflected from the upper and lower surfaces of the film. Newton's rings are formed in a similar way.

Interferometer. Any instrument or apparatus designed to produce interference fringes. Interferometers have a variety of uses including the accurate measurement of wavelength, the testing of flat surfaces, and the measurement of small changes in length.

Interferon. Protein produced by cells in the body in response to viruses. Interferon impairs the growth and replication of the attacking virus and has been shown to have some antitumor properties. Its possible role in the treatment of disease is still in the early stages of investigation.

Intermediate Vector Boson. Virtual particle.

Intermetallic Compound. A compound formed between two or more metallic elements combined in definite proportions. Intermetallic compounds are held together by metallic bonds and are often found as distinct phases in alloys.

Internal-Combustion Engine. Engine in which combustion of fuel takes place in a confined space, producing expanding gases that are used to provide mechanical power. The most common internal-combustion engine is the four-stroke reciprocating engine used in Automobiles. Here, mechanical power is supplied by a piston fitting inside a cylinder. A downstroke of the piston, the first stroke, draws a mixture of fuel and air into the cylinder from the Carburetor through an intake valve; the piston moves up to compress the mixture at the second stroke; at ignition, the third stroke, a spark from a spark plug ignites the mixture, forcing the piston down; in the exhaust stroke, an exhaust valve opens to vent the burned gas as the piston moves up. A rod connects the piston to a crankshaft. The reciprocating (up and down) movements of the piston rotate the crankshaft, which is connected by gearing to the drive wheels of the automobile. The ignition spark is provided by an electrical system whose power comes from a battery, which also supplies power to the starting system, a small electric motor that turns the crankshaft until the engine takes over. The engine is cooled by water circulating around the cylinders and is lubricated by motor oil driven around the moving engine parts by an oil pump. The two-stroke engine, which combines intake and compression in the first stroke and power and exhaust in the second, is used in lawn mowers and for small vehicles such as motorcycles. Other variations of the internal-combustion engine include the Wankel engine, which replaces the piston and cylinder with a triangular rotor moving inside an oval chamber, and the Diesel engine. (For a description of an external-combustion engine).

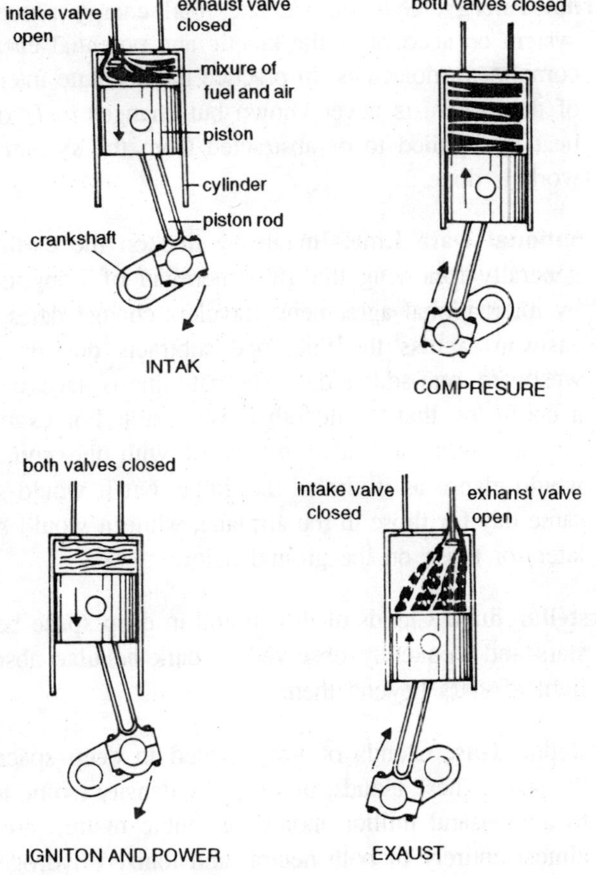

Fig. I-5. Internal-Combustion Engine.

Internal Conversion. A process in which an atomic nucleus in an excited state decays to its ground state and the energy released is used to eject an electron from an inner shell of the atom. The electron ejected is called a *conversion electron*. Internal conversion is a direct process resulting from coupling between electrons in the inner shells and the atomic nucleus. The ion produced, which is in an excited state, can subsequently emit an X-ray photon or an Auger electron.

Internal Energy. Symbol: U. The total energy possessed by a system on account of the kinetic and potential energies of its component molecules. In practice the absolute internal energy of the system is never known but changes in U occur when heat is supplied to or abstracted from the system and when work is done.

International Date Line. Imaginary line on the earth's surface, generally following the 180° meridian of Longitude, where, by international agreement, travelers change dates. Traveling eastward across the line, one subtracts one day; traveling westward, one adds a day. The date line is necessary to avoid a confusion that would otherwise result. For example, if an airplane were to travel westward with the sun, 24 hours would elapse as it circled the globe, but it would still be the same day for those in the airplane, while it would be one day later for those on the ground below.

Interstellar dust. Clouds of dust found in deep space between the stars and frequently observed as dark nebulae absorbing the light of stars beyond them.

Interstellar Gas. Clouds of gas situated in deep space between the stars. These clouds, of very low density (some ten million to a thousand million atoms per cubic metre), are made up almost entirely of both neutral and ionized hydrogen, though potassium, sodium, calcium, hydrocarbon compounds, and cyanogen have also been identified by their spectra.

Interstellar Matter. Matter in a galaxy between the stars. About

1% is tiny interstellar grains, believed to be frozen water vapor and carbon dioxide. The grains may appear optically as bright reflection or dark Nebulae. The rest of the matter is gas, mostly hydrogen and helium, but many molecules have been detected by Radio Astronomy techniques.

Interstitial. Pertaining to a hole formed between atoms in their positions in a crystal lattice. The atoms in a lattice can be thought of as spheres packed together in various ways and the interstices between these spheres are *interstitial positions.* Under favourable circumstances smaller atoms (*interstitial atoms*) can occupy these positions with little deformation of the original lattice. This type of structure is found in certain metal borides, carbides, and hydrides.

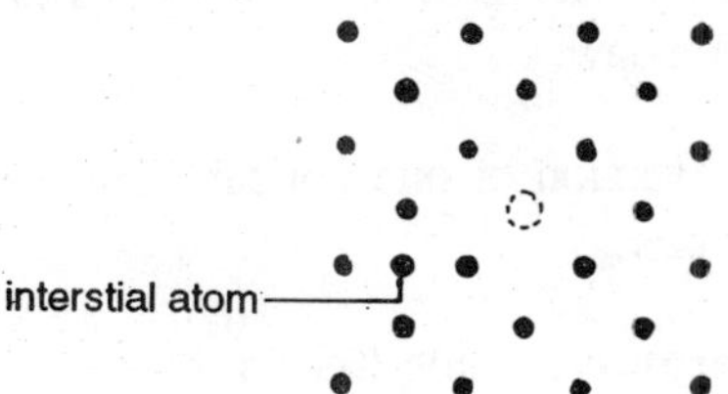

Fig. I-6. Frenkel defect.

A *Frenkel defect* is a type of defect in which an atom has moved from its normal position in the crystal to an interstitial position.

Intestine. Muscular, hoselike portion of the digestive System extending from the lower end of the Stomach to the anal opening. In humans the fairly narrow small intestine is a tubelike structure that winds compactly back and forth within the abdominal cavity. The contraction of its muscular walls (peristalsis, propels food onward while digestion is completed. Innumerable minute projections (villi) in the intestinal lining absorb the altered food for distribution by the Blood and Lymphatic Systems to the rest of the body. In the lower right abdominal cavity the small intestine joins the large intestine (colon), where most of the water content of the remaining mass is absorbed.

Intrauterine Device (IUD). Variously shaped Birth Control device, usually of plastic, which is inserted into the uterus by a physician. Apparently the IUD creates a hostile environment for the sperm or fertilized egg.

Intrinsic Semiconductor. Semiconductor.

Invar. An alloy of iron and nickel with a low coefficient of expansion, used in making pendulums, etc.

Inverse-square Law. Any law in which the magnitude of an effect is inversely proportional to the square of the distance from the cause. Examples are Coulomb's law, Amperes law, and Lambert's law.

Inversion. The conversion of one optical isomer of a compound to the other isomer.

Inversion Temperature. Joule-Kelvin effect.

Invertebrate. Any multicellular animal lacking a backbone. Invertebrates include all animals except the fishes, amphibians, reptiles, birds, and mammals, which are in the phylum Chordata. The major invertebrate groups are the segmented or Annelid Worms. Arthropods, Coelenterates, Echinoderms, Flatworms, Mollusks, Roundworms, and Sponges.

Invert Sugar. A mixture of equal parts of glucose and fructose made by hydrolysis of sucrose. The name refers to the fact that the mixture is laevorotatory whereas sucrose is dextrorotatory. Mutarotation occurs during hydrolysis.

In Vitro. Outside the organism: said of biochemical reactions performed in laboratory apparatus (literally "in glass").

In Vitro Fertilization. Technique for conception of a human embryo outside the mother's body. The ovum, or egg, is removed from the mother's body and placed in a special laboratory culture dish (Petri dish); sperm from the father are then added. If fertilization occurs, the fertilized ovum, after undergoing several cell divisions, is transferred to the mother's body for normal development in the uterus. First developed by Drs. Patrick C. Steptoe and Robert G. Edwards of Great Britain (where the first "test-tube baby" was born under their care in 1978), the technique was devised for use in cases of infertility when the woman's fallopian tubes are damaged and cannot be surgically repaired or the man's sperm count is low.

In vivo. In the natural living organism: said to biochemical reactions.

Involute. A curve that is the locus of the end of a string as it is unwound from another specified curve (the evolute).

Iodate. Any salt of iodic acid.

Iodide. Any of certain compounds containing iodine.

Iodine. I A bluish-black lustrous nonmetallic element found in several natural bromide salts that occur in sea water, Chile saltpetre, and brines from oil wells. Formerly obtained from seaweed, it is usually made by the action of chlorine on oil-well brine. Iodine has a violet irritating vapour and gives purple solutions with organic solvents. It is used as a germicide and in the manufacture of organic chemicals. Iodine belongs to the halogen group of elements. A.N. 53; A.W. 126.9045; m.pt. 113.5°C; b.pt. 184.4°C; r.d. 4.9; valency 1, 3, 5, or 7.

Iodine (I). Nonmetallic element, discovered in 1811 by Bernard Courtois. The least active of the Halogens, it is a dark-gray to purple-black solid that, when heated, sublimes directly to the vapor state. The violet-colored vapor has a characteristic irritating odor. Tincture of iodine and iodo-form have important medical uses. Silver iodide is used in photography. Starch turns deep-blue in the presence of iodine; this is a test for either substance. Thyroid-gland hormones contain iodine; inadequate dietary iodine results in goiter, a swelling of the thyroid. Thyroid defects are treated with iodine-131.

Iodoethane (ethyl iodide). A colourless oily liquid alkyl halide, C_2H_5I, made by reacting red phosphorus, iodine, and ethanol. M.pt. -108°C; b.pt. 72°C; r.d. 1.9.

Iodoform (tri-iodo methane). A greenish-yellow crystalline powder, CHI_3, with a characteristic penetrating odour. It is made by warming acetone with an alkaline iodine solution and is used as an antiseptic and anaesthetic. M.pt. 115°C; r.d. 4.1.

Iodomethane (methyl iodide). A colourless liquid alkyl halide, CH_3I, made by reaction between methanol and iodine in the presence of red phosphorus. M.pt. -66.45°C; b.pt. 42.4°C; r.d. 2.28.

Ion. An atom or group of atoms with a net positive or negative charge. Positive ions (*cations*) have a deficiency of electrons and negative ions (*anions*) have an excess. In formulae, the sign and charge of the ion is written as a superscript, as in Ca^{2+} which has a deficiency of two electrons.

Ion. Atom, or group of atoms, having a net electric charge, acquired by gaining or losing one or more electrons or protons. A simple ion consists of only one charged atom; a complex ion consists of an aggregate of atoms with a net charge. Because the electron and proton have equal but opposite until charges, the charge of an ion is always expressed as a whole number of positive or negative unit charges. If an atom or group loses electrons or gains protons, it will have a net positive charge and is called a cation. If an atom or group gains electrons or loses protons, it will have a net negative charge and is called an anion.

Ion Engine. An engine for propelling rockets by ejecting a stream of ions from the exhaust jet using a high electric field.

Ion Exchange. A process in which positive or negative ions are exchanged between a solution and a solid. Solid materials used in ion exchange are usually polymeric substances having charged sites to which ions are attached. Zeolites and other silicates, for instance, have a lattice of silicon and oxygen atoms with metal ions occupying certain positions in the lattice. If they are placed in contact with a solution ions, exchange occurs: ions in solution displace ions in the lattice; silicate ion exchanges are often used for removing the calcium ions from hard water and replacing them with sodium ions.

Ion-exchane resings are organic polymers containing ionized side groups. In cation exchange the side groups are carboxyl groups or sulphonic acid groups and positive metal ions are exchanged at these sites. In anion exchange the side groups are positive quaternary ammonium ions and negative ions are exchanged. Ion-exchange resins are extensively used for purifying and separating chemicals and for chromatographic analysis.

Ionic Bond. Electrovalent bond.

Ionic Crystal. Crystal.

Ion Implantation. Bombardment of a semiconductor with high-energy ions, which penetrate the surface, used as a method of controlled doping of transistors.

Ionization. The process of forming ions from neutral atoms or molecules. Ionization can occur as a result of several different processes. In solution, molecules may ionize spontaneously: $HCl = H^+ + Cl^-$. The energy necessary to break the bond is compensated by the solvation of the ions formed. Heterolytic fission can also occur in the gas phase as a result of the impact of particles or photons. Irradiation can also produce positive ions by ejecting one or more electrons: $A = A^+ + e$. Negative ions can be formed by electron capture: $A + e = A^-$.

Ionization Chamber. A chamber containing two electrodes in a gas with a potential difference between them, used to detect ionizing radiation by the current it causes in ionizing the gas between the electrodes.

Ionization Potential. symbol: I. The minimum energy that must be supplied to an atom or molecule to remove an electron (to infinity) and produce a positive ion. Ionization potentials are measured in electronvolts. The term *second ionization potential* is used in two ways. It is either the energy required to remove two electrons and form a doubly charged ion or it is the energy necessary to remove the second most strongly bound electron (from an inner shell or sub-shell) to form a singly charged ion in an excited state.

Ionizing Radiation. Electromagnetic radiation or particles that cause ionization. for this to occur the photon energy of the electromagnetic radiation or the kinetic energy of the particles must exceed the ionization potential of the substance ionized. The cross section for ionization must also be high. X-rays, ultraviolet radiation, and alpha particles are all forms of ionizing radiation: neutrons and gamma rays are much less effective.

Ionosphere. A region of ionized air and free electrons around the earth in the earth's upper atmosphere, extending from a height of about 50 km to 1000 km. It is produced by ionizing radiation, chiefly ultraviolet radiation, coming from the sun. The free electrons in the ionosphere reflect radio waves and the ionosphere can be used for radio transmissions between points that are hidden by the earth's curvature. It is divided into three distinct layers, differing in their concentrations of electrons. The layers are the *D-layer* (50-90 km), the *E-layer* or *Heaviside-Kennelley layer* (90-150 km), and the *F-layer* or *Appleton layer* (150-1000 km). The F-layer contains the highest electron concentration and is the one used for most radio transmission. At night, the concentration of electrons falls in the D-layer due to recombination of atoms and electrons.

Ionospheric Wave. Sky wave.

Ion Pump. A type of vacuum pump, used at very low pressures, in which gas is removed from the system by ionization and subsequent adsorption of the ions on a surface. The ions can be produced by electrons from a hot metal filament and adsorbed on a metal film. In one type of ion pump ionization occurs in a gas discharge between titanium electrodes. The discharge causes sputtering of the titanium, thus creating a continuously deposited film of titanium which acts as a getter. Such devices are called *getter-ion pumps* or *sputter-ion pumps.*

Iridium. Symbol: Ir. A silvery-white brittle transition element occurring associated with platinum. It is used in platinum-iridium and osmium-iridium alloys. The chemical properties of iridium are similar to those of rhodium. Its main compounds are iridium III and iridium IV complexes. A.N. 77; A.W. 192.2; m.pt. 2140°C; b.pt. 4130°C; r.d. 22.42; valency 1-4.

Iridium (Ir). Metallic element, discovered by Smithson Tennant in 1804. Chemically very unreactive, iridium is a very hard, usually brittle, extremely corrosion-resistant, silver-white metal used in chemical crucibles, surgical tools, and pivot bearings. The international kilogram standard is an iridium-platinum alloy.

Iris. 1. Eye.

2. An adjustable aperture for allowing light into an optical instrument.

Iris. Common name for members of the genus *Iris* of the Iridaceae, a family of perennial herbs that includes the Crocus, freesia (genus *Freesia*), and gladiolus (genus *Gladiolus*). The family is typified by modified stems (Bulbs, corms, and Rhizomes) and by linear or sword-shaped leaves. Distributed worldwide except in the coldest regions, the family is closely related to the Lily and Amaryllis families. The cultivated irises, freesias, and gladioli show a wide variety of colors in their usually perfumed blossoms. The many species of wild iris are often called flags. Orrisroot, a violet-scented flavoring used in dentrifices, perfumes, and other products, is made from the powdered rhizomes of several European species.

Fig. I-7. *Red iris,* Iris fulva.

Iron. Symbol: Fe. A silvery-white soft malleable and ductile ferromagnetic transition element; the fourth most abundant element in the earth's crust, occurring in many minerals, the main ores being haematite (Fe_2O_3), limonite [$FeO(OH).nH_2O$], and magnetite (Fe_3O_4). Iron makes up most of the earth's core and is also found in some meterorites. It is obtained by smelting in a blast furnace to give pig iron, which is then converted into cast iron, wrought iron, or steel. The pure element is rarely prepared. It exists in several allotropic forms.

Chemically, the element is reactive, being quickly oxidized in moist air and combining directly with most nonmetals on heating. It dissolves in dilute nonoxidizing acids. Iron forms two series of compounds: iron II (*ferrous*) compounds and iron III (*ferric*) compounds. The element forms numerous ionic salts in both these oxidation states as well as many complexes. Iron also forms compounds in other oxidation states, notably the carbonyls and ferrates. A.N. 26; A.W. 55.847; m.pt. 1535°C; b.pt. 2750°C; r.d. 7.87; valency 2 ,3, 4, or 6.

Iron (Fe). Metallic element. Iron is a lustrous, ductile, malleable, silver-gray metal. A good conductor of heat and electricity, it is attracted by a magnet and is itself easily magnetized. Iron is abundant in the universe; it is found in many stars, including the sun. It is the fourth most common element in the earth's crust, of which it constitutes about 5% by weight, and is believed to be the main component of the earth's core. Iron is rarely found uncombined in nature except in Meteorites, and it rusts readily in moist air. Limonite and hematite are the chief iron ores; other ores include Magnetite, taconite, and siderite. Iron ores are refined in a Blast Furnace to produce pig iron, which can be remelted and poured into molds to make Cast Iron, commercially purified to make Wrought Iron, or alloyed with carbon and other elements to make Steel. Iron compounds are used as paint pigments, in dyeing, and in ink manufacture. A component of Hemoglobin, the oxygen-carrying pigment of red Blood cells in vertebrates, iron is important in Nutrition; one cause of Anemia is iron deficiency.

Iron Age. Technological period from the first general use of iron to modern times. In Asia, Egypt, and Europe it followed the Bronze Age; Europeans brought it to the Americas. Hammered iron was known in Egypt before 1350 B.C., and after the fall of the Hittite empire (1200 B.C.) migrants carried iron technology through S. Europe and the Middle East. Smelting was known in the Etruscan Civilization. After the Early Iron Age (C.800-500) in central Europe, migrating Celts spread the use of iron into W. Europe and the British Isles, where Late Iron Age peoples used it for tools, vehicles, coinage, and art objects. Metal technology increasingly influenced Culture up to the industrial Revolution and beyond.

Iron Chloride. Either of two chlorides of iron. *Iron II chloride* (ferrous chloride) is a white crystalline solid, $FeCl_2$ or $FeCl_2.4H_2O$, used as a mordant. M.pt. 670°C; r.d. 3.2. *Iron III chloride* (ferric chloride) is an orange crystalline solid, $FeCl_3.6H_2O$, used in the treatment of sewage and industrial waste. It is black when anhydrous. M.pt. 300°C; r.d. 2.9.

Iron Oxide. Any of three oxides of iron. *Iron II oxide* (ferrous oxide) is a black powder, FeO, obtained by heating iron II oxalate. M.pt. 1420°C; r.d. 5.7. *Tri-iron tetroxide* (feeosoferric oxide) is a red or black ferromagnetic powder, Fe_3O_4, made by reacting hot iron with air or steam and used in pigments and polishes. M.pt. 1538°C (decomposes); r.d. 5.2. The highest oxide, *iron III oxide* (ferric oxide), occurs naturally as haematite. It is a red powder, Fe_2O_3, used as a pigment and as a metal polish (*jeweller's rouge*). M.pt. 1565°C; r.d. 5.1.

Iron Sulphate. Either of two salts of iron. *Iron II sulphate* (ferrous sulphate) is a green crystalline solid, $FeSO_4.7H_2O$, prepared by the action of sulphuric acid on iron, and used for making other iron salts, sewage treatment, and as a dietary supplement. It is also known as *green vitriol* and *copperas*. M.pt. 64°C; r.d. 1.9. *Iron III sulphate* (ferric sulphate) is a pale yellow powder, $Fe_2(SO_4)_3$, used in analytic chemistry. R.d. 3.1.

Irradiation. Exposure to radiation, usually with the aim of inducing some change as a result of the impact of particles or photons.

Irreversible. Reversible.

Irreversible Reaction. A chemical reaction in which the reactants are completely converted into the products.

Irrigation. In Agriculture, artificial watering of the land, using water diverted from rivers and lakes or pumped from underground. Modern large-scale irrigation is often part of a multipurpose project that may also produce hydroelectric power and provide systems for water supply and flood control.

Surface irrigation delivers water to a field directly from a canal, well, or ditch. In surface-pipe irrigation, the water is piped to the field and distributed via-sprinklers or smaller pipes. Drip irrigation, a recent development, delivers a small, measured amount of water to each plant through narrow plastic tubes and helps prevent the two major problems caused by irrigation: waterlogging, saturation of the soil as a result of inadequate drainage; and soil salinization, the accumulation of salts deposited by irrigation water.

Isobar. 1. A curve or equation relating quantities measured at the same pressure. An example is a graph of the volume of a gas against its temperature, with the pressure kept constant.

2. Any line joining points that have equal pressures.

3. Any of two or more isotopes that have different atomic numbers but the same mass numbers. For example, the isotopes palladium-105 and silver-105 are isobars. Their atomic numbers are 46 and 47 respectively and their atomic masses are almost identical.

Isobutyl Group. Butyl group.

Isochore. A curve or equation relating quantities measured at the same volume. An example is Van't Hoff's isochore.

Isocline. A line on the earth connecting points that have the same magnetic dip.

Isocyanate. Cyanic acid.

Isocyanic Acid. Cyanic Acid.

Isodiapheres. Nuclides that have the same difference between total number of neutrons and total number of protons. Examples are 190/78 Pt and 186/76 Os, the difference being 34 in each case.

Isodynamic Line. A line on the earth connecting points that have the same horizontal intensity of magnetic field.

Isoelectronic. Denoting atoms or molecules that have the same numbers of electrons. Carbon dioxide and nitrous oxide, for example, have isoelectronic molecules.

Isogonal Line. A line on the earth connecting points that have the same magnetic declination.

Isomer. In chemistry, one of two or more compounds having the same molecular formula (*i.e.*, the same number of atoms of each element in a molecule) but different structures (arrangements of atoms in the molecule). Isomers have the same number of atoms of each element in them and the same atomic weight but differ in other properties. *Structural isomers, e.g.*, Ethanol (CH_3CH_2OH) and dimethyl ether (CH_3OCH_3), differ in the way the atoms are joined together in their molecules. Stereoisomers have the same basic arrangement of atoms in their molecules but differ in the way the atoms are arranged in space. Geometric isomers, which are stereoisomers that differ in the positioning of groups about a double bond or some other feature that gives the molecule a certain amount of structural rigidity, differ in physical properties such as melting and boiling points. Optical isomers are stereoisomers in which the two molecules are mirror images of each other and, each being asymmetrical, cannot be superposed on each other, optical isomers differ in the direction in which they are rotate light passed through the molecules.

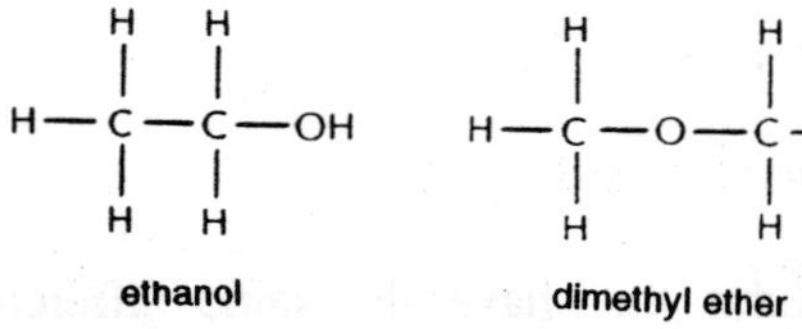

```
    H   H                 H         H
    |   |                 |         |
H — C — C — OH        H — C — O — C — H
    |   |                 |         |
    H   H                 H         H

   ethanol              dimethyl ether
```

Fig. I-8. Isomer.

Isomerism. 1. The existence of two or more chemical compounds with the same molecular formula but different arrangements of atoms in their molecules. Such compounds are said to be *isomers*. Several types of isomerism are distinguished, broadly classified into structural isomerism and stereoisomerism.

2. The existence of nuclei with the same atomic number and mass number but with different energy states. A nucleus in its ground state and one in an excited state constitute a pair of nuclear *isomers*.

Isometric. Cubic.

Isomorphism. The occurrence of two or more different substances with the same crystal structure; the substances are said to be *isomorphous*. The term is often used in a more restricted sense with reference to compounds that have similar compositions and can crystallize together in mixed crystals.

Isopentane (2-methylbutane). A colourless flammable liquid alkane, $CH_3CH\ (CH_3)C_2H_5$, obtained from petroleum. It is isomeric with pentane and neopentaane. M.pt. -159.9°C; b.pt. 279°C; r.d. 0.62.

Isopentyl Group. Pentyl group.

Isoprene. A colourless volatile liquid, $CH_2{:}C(CH_3)CH{:}CH_2$, made from refinery and coal gases and tars. It polymerizes readily and is used in making synthetic rubber. Isoprene is the structural unit of terpenes. M.pt. -146°C; b.pt. 34.1°C; r.d. 0.68.

Isoscelese Triangle. A triangle with two sides equal.

Isospin (isotopic spin). A property of elementary particles used to characterize particles in the same multiplet. The proton and neutron, for example, differ in charge but are identical in their strong interactions. They can be regarded as two states of the same particle. The property is described by quantum numbers in a similar way to spin. The multiplet is characterized by the *isospin* (symbol: *I*) which has values 0, 1/2, 1. etc. The particles within a multiplet have different values of *isospin quantum number* (symbol: I_3) equal to $-I$, $-(I-1)$, $(I-1)$, I. Thus in the proton-neutron multiplet, in which I = 1/2, the proton has I_3 = 1/2 and the neutron has I_3 = −1/2.

Isotactic. Polymer.

Isotherm. An equation or a curve on a graph representing behaviour of a system at a constant temperature.

Isothermal. Denoting a process in which no change occurs in the temperature of the system. In an isothermal change the system is always in thermal equilibrium with its surroundings. *Compare* adiabatic.

Isotones. Nuclides that have the same neutron number but different atomic numbers.

Isotonic. Denoting solutions that have the same osmotic pressure.

Isotope. Any of a number of forms of an element, all of which differ only in the number of neutrons in their atomic nuclei. Two isotopes of the same element thus have nuclei with the same atomic number and different mass numbers. Carbon, for instance, is a mixture of two isotopes, one with six protons and six neutrons and the other with six protons and seven neutrons. The respective percentages of carbon-12 and carbon-13 are 98.892 and 1.108, leading to an overall atomic weight of carbon of 12.011. All elements also have artificial isotopes produced by bombardment with high-energy particles. Isotopes of an element have identical chemistries because they contain the same number of electrons. They do, however, have different physical properties and chemical reactions involving different isotopes take place at slightly different rates. These two properties are used in separating isotopes. The principle methods of separation are diffusion, thermal diffusion, centrifugal separation, electrolytic separation, the use of kinetic isotope effects, and the use of large mass spectrometers. The last method is used only for very small samples.

Isotope. One of two or more atoms having the same Atomic Number but differing in Atomic Weight and Mass Number. The nuclei of isotopes of the same element have the same number of Protons (equal to the element's atomic number) but have different numbers of Neutrons. The isotopes of a given element have identical chemical properties but slightly different physical properties. A radioactive isotope, or radioisotope, is a natural or artificially created isotope having an unstable nucleus that decays emitting alpha, beta, or gamma rays until stability is reached. For most elements, stable and radioactive isotopes are known.

Isotopic Number (neutron excess). The difference between the number of neutrons and the number of protons in a nucleus.

Isotopic spin. Isospin.

Isotopic Weight. The atomic weight of an isotope. The isotopic weight is almost the same as the mass number.

Isotropic. Having properties that are independent of direction. For example, an isotropic crystal has the same physical properties (refractive index, modulus of elasticity, etc.) along all its axes because of its crystal structure. *Compare* anisotropic.

Ivory. Type of dentin present only in Elephant tusks. Major sources are Africa and Asia. Ivory was long used for carvings and inlay works and as a surface for miniature paintings. The ancient civilizations of Egypt, Assyria, Babylon, Greece, Rome, India, China, Japan, Byzantium, and early Christian Europe produced works in ivory. In modern times it has been used chiefly to make keys for pianos and other instruments, billiard balls, and handles. The diminishing elephant population (largely the result of their slaughter for ivory) and efforts to protect existing herds have led to greater use of substitutes, such as Plastic. In the past, teeth or tusks of such animals as the hippopotamus, walrus, narwhal, sperm whale, and wild boar were also called ivory.

Ivy. Name applied loosely to any trailing or climbing plant, particularly cultivated forms; more properly, the English ivy (*Hedera helix*) and some other members of the family Araliaceae. The evergreen English ivy, grown in many varieties, is a popular ornamental vine, its berries are poisonous. The Grape family also includes some ivies, most notably the Virginia Creeper and Boston ivy. Kenilworth ivy (*Cymbalaria muralis*), of the figwort family, is often used as ground cover.

Jackal. Carnivorous Mammal (genus *Canis*) related to the Dog and Wolf and similar in size and behavior to the Coyote. Some authorities classify jackals in a separate genus (*Thos*). Jackals are found in Africa and S Asia, where they inhabit deserts, grasslands, and bush country. They forage by night and spend the day in holes or bush.

Jade. Common name for either of two minerals, both white to green in color, used as Gems. Jadeite [$NaAI(SiO_3)_2$], rarer and costlier,is found in Burma, China, Japan and Guatemala. Nephrite [$Ca_2(Mg,Fe)_5Si_8O_{22}(OH,F)_2$] occurs in New Zealand, Central Asia, Siberia, and parts of North America. Jade was much used by primitive peoples to make implements and has been prized by the Chinese and Japanese as the most precious of gems.

Jaeger's method. An experimental method for measuring surface tension by determining the pressure required to produce a bubble of air from a capillary tube immersed in the sample liquid.

Jaguar. Large Cat (*Panthera onca*) found from the SW U.S. to S central Argentina. It has a yellow or tawny coat with black rings and spots; some rings surround spots, a feature which distinguishes it from a Leopard. An adult male may be $4^1/_2$ ft (1.4 m) long (excluding the long tail), stand $2^1/_2$ ft (76 cm) high at the shoulder, and weigh 200 lb (90 kg). Mainly forest dwellers, jaguars are also found in rocky, semidesert areas and on the Pampas.

Jasmine or **Jessamine.** Plant (genus *Jasminum*) of the Olive family, chiefly of Old World tropical and subtropical regions but cultivated in other mild areas and greenhouses. The blossoms, mostly yellow or white and usually fragrant, are used in scenting tea; the oil is used in perfumery.

Jaspers, Karl. 1883-1969, German philosopher. Generally placed within the orbit of Existentialism, Jaspers believed that genuine philosophy must spring from the study of a person's individual existence, which he viewed as enclosed by an all-embracing, trancendental reality he called "the encompassing." Among his works are *Man in the Modern Age* (1931) and *Philosophy* (3 vol., 1932).

Jaundice. Abnormal condition in which body fluids and tissues, particularly the skin and eyes,take on a yellowish color as a result of excess bilirubin, a substance normally removed form the bloodstream by the Liver and eliminated from the body in Bile. Causes include excessive disintegration of red blood cells (as in some types of Anemia); damage to liver cells by disease (e.g., Cirrhosis, Hepatitis); and bile duct blockage (e.g., by gallstones or tumors). Jaundice in the newborn may indicate an Rh Factor reaction.

Jay. Bird (family Corvidae) related to the Crow and found in Europe, Asia, and the Americas. The blue jay (*Cyanocitta cristata*) of central and E. North America has a grayish blue crest and upper parts, and bright blue wings with white and black markings. The Canada jay (*Perisoreus canadenis*), gray with a white throat and forehead and no crest, is found in northern coniferous forests and swamps.

Jellyfish. Free-swimming state (see Polyp and Medusa) of invertebrate animals (Coelenterates) of the classes Hydrozoa and Scyphozoa of the phylum Cnidaria. Many jellyfish are colored and are considered to be among the most beautiful of animals. They have bell- or umbrella-shaped bodies with a clear, jellylike material filling most of the space between the upper and lower surfaces. A mouth is located on the underside and tentacles dangle from the bell margin. Most catch their prey with stinging cells located in the tentacles and most are marine, living in ocean depths and along the coasts.

Jenner, Edward, 1749-1823. English physician. His experiments, beginning in 1796 with the vaccination of James Phipps, proved that cowpox provided immunity against smallpox, helped rid many areas of the disease, and laid the foundations of modern immunology as a science.

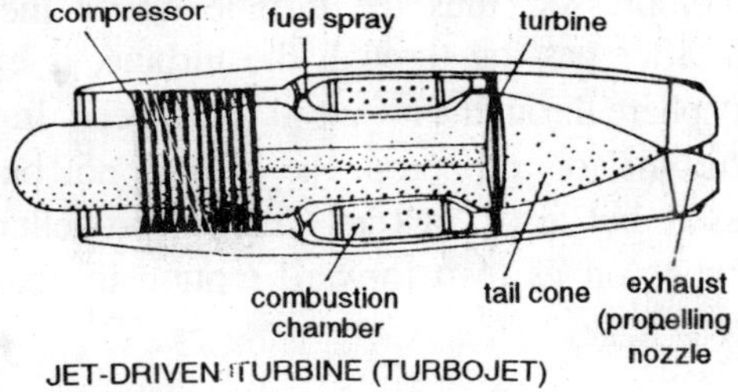

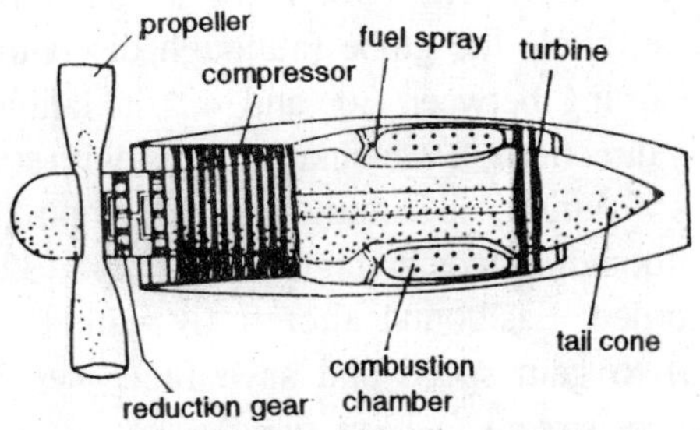

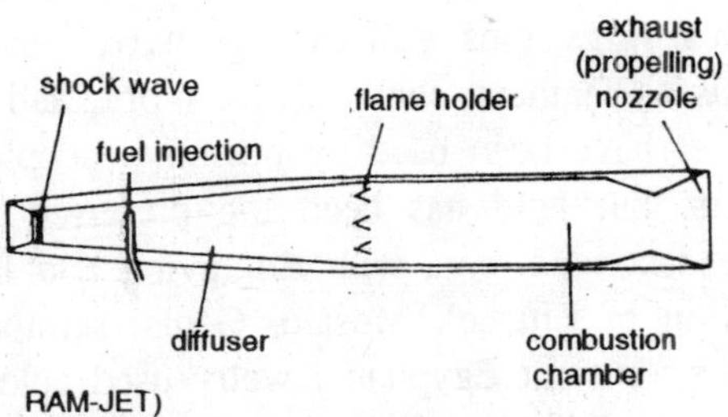

Fig. J-1. Jet propulsion engines.

Jet propulsion. Propulsion of a body, such as an airplane, by a forced developed in reaction to the ejection of highspeed jet of gas. The jet-driven turbine, or turbojet, consists of four basic parts: compressor, combustion chambers, Turbine, and propelling nozzles. In the combustion chambers the combustion of a fuel, mixed with air coming from the compressor, generates expanding gases, which spin the rotor of the turbine. The axis of the turbine is connected by a shaft directly to the axis of the compressor; thus the turbine drives the compressor. The gas, after passing through the turbine, is exhausted into the atmosphere through a nozzle at high speed. In the propeller-driven turbine, or turboprop, the turbine not only drives the compressor, but also turns an outside propeller. The ramjet engine relies on its own forward motion to compress the air that enters it.

Jet stream. Narrow, swift air currents at altitudes of 7 to 8 mi (11.3 to 12.9 km). The two major jet streams, one in each hemisphere, circle the globe (although discontinuous at some points), varying between 30° and 40° in latitude. They flow in easterly directions in wavelike patterns, with speeds averaging 35 mph (56 km/hr) in summer and 75 mph (120 km/hr) in winter, although speeds as high as 200 mph (320 km/hr) have been recorded. Eastbound aircraft fly with the jet stream (a tail wind) to gain speed and save fuel; west-bound aircraft avoid the jet stream (a head wind).

Jewelry. Personal adornments worn for ornament, to show rank or wealth, or to follow custom. Jewelry predates clothing, and necklaces, rings, pins, and earrings have been made of such materials as feathers, shells, teeth, ivory, and metal. Bronze and silver have been used by primitive peoples and modern designers, but gold has been the preferred metal. Jewelry may be decorated, e.g., with Engraving and Etching and by application of Enamel, Mosaic, Gems, semiprecious stones, and glass. Ancient Egyptian jewelry used colorful beads and emblematic motifs, e.g., the scarab. The Greeks made ornaments of plain gold; after 400 B.C. gems and, later, Cameos were used. Roman jewelry was massive, and ropes of pearls were prized. Medieval jewelry included large brooches and heavy girdles. In the Renaissance, men and women wore gold chains, jeweled collars,and pendants, often designed or even rendered by noted artists. Ornaments crowded with stones were worn to excess in the late 17th cent. Jewelry was almost superseded in the late 18th cent. by decorative buttons, watches, and snuffboxes, but the 19th cent. saw the revival of the bracelet and cameo. With factory production of jewelry, artistry declined. In the 20th cent. costume jewelry was introduced by Gabrielle Chanel (see Fashion, table). Since the craft revival of the 1960s there has been renewed interest in hand-wrought jewelry.

Jewelweed. Common name for the Balsaminaceae, a family of widely cultivated annual and perennial herbs. The principal genus is *Impatiens,* found in tropical and north temperate regions and so named because its seed capsules burst when touched; some species are also commonly called touch-me-not. A few species are grown as ornamentals, e.g., the garden balsam (*I. balsamina*).

Joint. In anatomy, juncture between two bones. Some joints are immovable, e.g., those connecting bones of the skull. Hinge joints provide a forward and backward motion, as at the elbow and knee. Pivot joints permit rotary movement, like the turning of the head from side to side. Ball-and-socket joints, like those at the hip and shoulder, allow the greatest range of movement. Ease of movement is aided by elastic cartilage, lubricating synovial fluid, and, in some joints, a cushioning, fluid-filled sac (bursa), which reduces friction.

Joly's steam calorimeter. An apparatus for measuring the specific heat capacities of gases at constant volume. It consists essentially of two equal metal spheres suspended from opposite arms of a balance, one sphere being evacuated and the other filled with a known volume of the sample gas. The spheres are contained in an enclosure through which steam is passed: the specific heat capacity may be calculated by measuring the difference between the amounts of water condensing on the spheres.

Joule. Symbol: J. The SI unit of energy equal to the work done when the point of application of a force of one newton is moved one metre, movement being in the direction of the force. It is the work done by a current of one ampere flowing through a resistance of one ohm for one second.

Joule heating. The heating of a conductor by an electric current flowing through it. The heat produced is I^2Rt, where I is a steady current flowing for a time t through a resistance R.

Joule, James prescott. 1818-89, English physicist. He established the mechanical theory of Heat and was the first to determine the relationship between heat energy and mechanical energy (the mechanical equivalent of heat). Joule discovered the first law of Thermodynamics, which is a form of the law of conservation of Energy. The mechanical unit of Work, the joule, is named for him.

Joule-Kelvin effect. A change in temperature occuring in a gas when it undergoes an adiabatic expansion. The effect can be demonstrated by allowing the gas to flow through a porous plug from a region of high pressure into a region of low pressure. Usually a slight fall in temperature is observed, caused by the fact that, in expanding, the gases have to overcome forces between their molecules. Substances have a characteristic temperature—the *inversion temperature*—above which they are heated by expansion and below which they are cooled.

Junction transistor. *See* transistor.

Juniper. Aromatic evergreen tree or shrub (genus *Juniperus*) of the Cypress family, widely distributed over the north temperate zone. Many are important sources of lumber and oil. The insect-repellent wood of the the red cedar (*J. virginiana*) is especially valuable for closets, furniture, and posts; the oil is used in medicine and perfumery. The common juniper (*J. communis*) and its varieties (e.g., dwarf and pyramidal) are grown ornamentally; its fruits are used to flavor Gin.

Jupiter. The fifth planet in succession outwards from the sun and the first of the giant planets. Lying beyond the asteroid belt at a distance of 778 million kilometres from the sun, Jupiter is by far the largest of the planets, with an equatorial diameter of 142 800 km and a mass 318 times that of the earth. Despite these dimensions, the planet's relative density is only 1.334—a fifth the size of the earth's. Its gravity is 2.5 times that of the earth.

Jupiter is a typical giant planet, having a dense impenetrable atmosphere of methane, ammonia, and hydrogen. The rotation is not uniform, being slowest at the equator and increasing towards the poles: an average value of 9 hours 55 minutes can be assigned. Jupiter has several interesting features. It has a magnetic field 30 times as strong as the earth's field. It also shows a vast marking in the southern hemisphere—the *Great Red Spot.* While other optical features of Jupiter are transient, this phenomenon, whose origin remains unknown, has been present since serious telelscope observation began. Jupiter has 12 satellites.

Jupiter. In astronomy, 5th Planet from the sun, at a mean distance of 483.6 million mi (778.3 million km), and largest planet in the solar system, with an equatorial diameter of 89,400 mi (143,800 km). It is a gaseous planet with an atmosphere composed mostly of hydrogen and helium, with traces of methane, ammonia, and other gases, and about five or six zones each of counterflowing eastward- and westward-flowing winds. The most prominent atmospheric features (all in the southern hemisphere) are the Great Red Spot, at least 300 years old and measuring c.30,000 by 10,000 mi (48,000 by

16,000 km), and three large white ovals that formed in 1939. Jupiter has 16 known natural satellites, 12 of which are of small diameter. The four larger Galilean satellites were discovered by Galileo in 1610. *Callisto* (diameter: 2,995 mi/ 4,820 km), the most distant and the least active geologically of the four, has a heavily cratered surface. *Ganymede* (diameter: 3,279 mi/5,276 km), second most distant of the four and the largest satellite in the solar system, has heavily cratered regions, tens of miles across, that are surrounded by younger, grooved terrain. *Europa* (diameter: 1,942mi/3,126 km) is a white, highly reflecting body whose smooth surface is entirely covered with dark streaks up to 70 km in width and from several hundred to several thousand kilometers in length. *Io* (diameter: 2,257 mi/3,632 km), the closest to Jupiter of the four, is the most active geologically, with eight active volcanoes that are probably energized by the tidal effects of Jupiter's enormous mass. The red color of *Amalthea* (diameter: 150 mi/240 km), a small, elongated satellite interior to Io's orbit that was discovered (1892) by Edward Barnard, probably results from a coating of sulfur particles ejected from Io. Four Space Probes have encountered the Jovian system: *Pioneers 10 and 11* (1973 and 1974) and *Voyagers 1 and 2* (both 1979). The latter two discovered Io's volcanoes, a thin ring system surrounding Jupiter, and three of the smaller satellites. The U.S. space probe *Galileo* will orbit Jupiter in the late 1980s and send probe into the planet's atmosphere.

Jupiter. In Roman mythology, supreme god; also called Jove; son of Saturn and Ops; brother and husband of Juno. Originally an agricultural god, he developed into the prime protector of the state and was identified with the Greek Zeus.

Jute. Tropical annual (genus *Corchorus*) of the Linden family, and its fiber. Although the fiber, of comparatively low Cellulose content, is weak and deteriorates quickly, it is the principal coarse fiber in commercial production; Chief sources are *C. capsularis* and *C. olitorius*. India, with its hot climate and abundant cheap labor, is the unrivaled world producer and processor. Easily dyed and spun, jute is used for coarse fabrics, especially burlap and sacking, and for twine, rope, and insulation.

Kainite. A mineral consisting of a hydrated mixed salt of magnesium sulphate and potassium chloride, $MgSO_4$–$KCl.3H_2O$, used as a source of potassium fertilizers.

Kale, borecole, and **collards.** Common names for nonheading types of Cabbage (mostly of the variety *acephala*), cool-weather crops of Mustard family. They are grown for their edible greens and, in Europe, for fodder.

Kangaroo. Hopping Marsupial of the family Macropodidae, found in Australia, Tasmania, and New Guinea. Knagaroos have powerful hind legs, long feet, short forelimbs, and long, muscular tail for maintaining balance. The female has a pouch in which she suckles the newborn kangaroo (joey). The chief grazes of Australian plains, kangaroos are diurnal and live in herds. Great red kangaroo males may attain a height of 7 ft (210 cm) and weigh over 200 lb (90 kg). Smaller members of the kangaroo family include the rabbit-sized wallabies and pademelons. Kangaroos have been extensively hunted as pests and for their valuable hides.

Kant, Immanuel. 1724-1804. German philosopher, one of the greatest figures in the history of Metaphysics. After 1755 he taught at the Univ. of Konigsberg and achieved wide renown through his teachings and writings. According to Kant, his reading of Hume woke him from his dogmatic slumber and led him to become the "critical philosopher," synthesizing the rationalism of Leibniz and the Skepticism of Hume. Kant proposed that objective reality is known only insofar as it conforms to the essential structure of the knowing mind.

Only objects of experience, phenomena, may be known, whereas things lying beyond experience, noumena, are unknowable even though in some cases we assume a priori knowledge of them. The existence of such unknowable "things-in-themselves" can be neither confirmed nor denied, nor can they be scientifically demonstrated. Therefore, as Kant showed in the *Critique of Pure Reason* (1781), the great problems of metaphysics—the existence of God, freedom, and immortality—are insoluble by scientific thought. Yet he went on to state in the *Critique of Practical Reason* (1788) that morality requires belief in their existence. Kant's Ethics centers in his categorical imperative, or absolute moral law, "Act as if the maxim from which you act were to become through your will a universal law." His *Critique of Judgment* (1790) considered the concepts of beauty and purposiveness as a bridge between the sensible and the intelligible worlds. Kant's influence on modern philosophy has continued to the present day. His work fostered the development of German Idealism by Fichte, Schelling, and Hegel. The Neo-Kantianism of the late 19th cent. applied his insights to the study of the physical sciences (Hermann Cohen, Ernst Cassirer), and to the historical and cultural sciences (Heinrich Rickert); his influence is also seen in the thought of Dilthey; in the pragmatism of Dewey and William Jams; in the theology of Schleiermacher; and in Gestalt psychology.

Kaon (K-meson). A type of meson having either zero charge and a mass 974.2 times that of the electron, or a positive or negative charge and a mass 966.6 times that of the electron.

Kater's pendulum. A pendulum designed for the accurate measurement of the acceleration of free fall. It consists of a metal bar with knife edges mounted near each end and two weights that can be slid up and down the bar between the knife edges. The pendulum is pivoted on each set of knife edges in turn and the positions of the weights are adjusted until the period of oscillation is the same for each pivot. Under these conditions the period is the same as that of a simple pendulum ($2\pi(l/g)^{1/2}$), with a length l equal to the distance between the pivots. [After Henry Kater (1777-1835), British physicist.]

Keeper (armature). A small piece of iron or steel placed across the poles of a permanent magnet when it is out of use to complete the magnetic circuit and prevent loss of magnetism.

Kekule formula. See aromaticity. [After F.A. Kekule von Stradonitz (1829-96), German chemist.]

Kekule von Stradonitz, Friedrich August (ka'koola), 1829-96. German organic chemist. A professor at Bonn, he is most noted for his representation of the molecular structure of Benzene as a ring. Such rings are the basic structural feature of Aromatic Compounds.

Kelvin. Symbol : K. The basic SI unit of thermodynamic temperature, equal to1/273.16 of the thermodynamic temperature of the triple point of water. This unit was formerly the degree Kelvin (°K). It is equal to one degree on the centigrade scale of temperature. [After Lord Kelvin (William Thomson) (1824-1907), British physicist.]

Kelvin, William Thomson, 1st **Baron,** 1824-1907. British mathematician and physicist; b. Ireland. He was professor (1846-99) of natural philosophy at the Univ. of Glasgow. His work in Thermodynamics coordinating the various existing theories of heat established the law of the conservation of Energy as proposed by James Joule. He discovered the Thomson effect in thermoelectricity and introduced the Kelvin scale, or absolute scale, of Temperature. His work on the transmission of messages by undersea cables made him a leading authority in this field.

Kennelley layer. *See* ionosphere.

Kepler, Johannes, 1571-1630. German astronomer. He was professor of mathematics at Graz (1593-98) and court mathematician to Holy Roman Emperor Rudolf II. In 1596 he wrote *Mysterium cosmographicum,* which led to exchanges with Galileo and Tycho Brahe. His *Astronomia nova* (1609) contained the first two of what became Kelper's Law; the third law appeared in 1619 in his *Harmonice mundi.* These laws were the result of calculations based on Brahe's accurate observations, which Kepler published in the *Tabulae Rudolphinae* (1627).

Kepler's laws. Three mathematical statements made by Johannes Kepler that accurately describe the revolutions of the planets around the sun. The first law states that the shape of each planet's orbit is an ellipse (see Conic Section) with the sun at one focus. The second law states that if an imaginary line is drawn from the sun to the planet, the line will sweep out equal areas in space in equal periods of time for all points in the orbit. The third law states that the ratio of the cube of the semimajor axis of ellipse (i.e., the average distance of the planet from the sun) to the square of the planet's period (the time it needs to complete one revolution around the sun) is the same for the planets. Newton give a physical explanation of Kepler's laws with his laws of Motion and law of Gravitation.

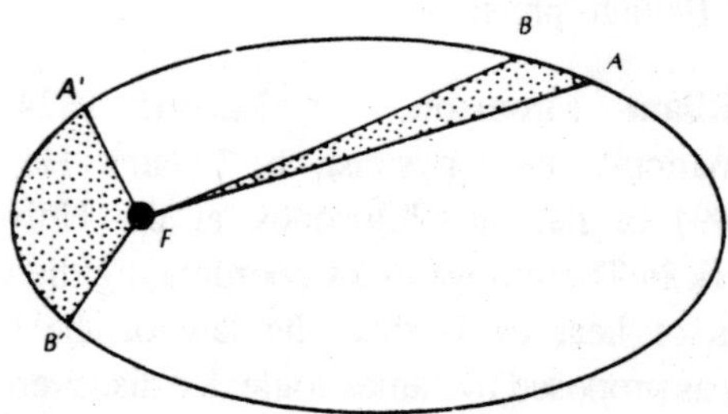

Fig. K-1. Kepler's law: Schematic representation of Kepler's second law. The area ABF and A'B'F are equal and are swept out in equal intervals of time by a planet orbiting around the sun (at F).

Kepler's laws. Three laws of planetary motion :

1. The planets move in elliptical orbits with the sun at one of the foci of the ellipse.

2. The radius vector (the line joining the centre of the planet to the centre of the sun) sweeps out equal areas of space in the same interval of time.

3. The square of a planet's sidereal period is proportional to the cube of its average distance from the sun.

[Formulated between 1609 and 1619 by Johannes Kepler (1571-1630), German astronomer.]

Kepler telescope. A type of refracting telescope consisting of a convex objective lens and a convex eyepiece. The lenses are separated by a distance equal to the sum of their focal lengths and the magnification is the ratio of the focal lengths. The instrument produces an inverted image. It can be modified for use as a terrestrial telescope by inclusion of an erecting lens or prism.

Kerosene or **Kerosine.** A colourless, thin oil that is less dense than water. It is a mixture of Hydrocarbons commonly obtained in the fractional Distillation of Petroleum, but also from coal, oil shale, and wood. Once the most important refinery product because of its use in lamps, kerosene is now used chiefly as a carrier in insecticide sprays and as a fuel in jet engine.

Kerr effect. 1. The phenomenon in which some liquids or gases show double refraction in the presence of a strong electric field. The field is applied at right angles to the direction of the light. A *Kerr cell* is a transparent cell containing a suitable liquid and two electrodes. A beam of plane-polarized light passing through the cell can be stopped by applying a voltage across the electrodes : the devices acts as a shutter.

2. The phenomenon in which a beam of plane-polarized light is elliptically polarized by reflection from the polished pole face of an electromagnet. [After John Kerr (1824-1907), British physicist.]

Ketal. Any of a class of organic compounds that have the general formula RR'C (OR'') (OR'''), where R indicates a hydrocarbon group. Ketals are formed in the same way as acetals by reaction of a ketone with an alcohol.

Ketene (ethenone). A colourless toxic gas, $CH_2 : C : O$, with a penetrating odour, made by pyrolysis of acetone. It is an acetylating agent, used in the manufacture of aspirin and cellulose acetate. M.pt.-150°C; b.pt. - 56°C.

Keto-enol tautomerism. A form of tautomerism in which transfer of a hydrogen atom occurs from a carbon atom next to a carbonyl group to the oxygen atom of the carbonyl group, thus producing an unsaturated alcohol. The two isomeric forms of the compound are called the *keto* and *enol* forms of the compound.

Keto form. *See* keto-enol tautomerism.

Ketone. Any of a class of organic compounds with the general formula RCOR', where R and R' are usually hydrocarbon groups. Ketones can be prepared by oxidation of secondary alcohols. They give a positive reaction with Fehling's solution. They also have a variety of addition and condensation reactions.

Ketose. A sugar that is a ketone in its straight-chain form. *Compare* aldose.

KeV. Kiloelectronvolts : 1000 electronvolts.

Kidney. One of a pair of small, bean-shaped organs of the Urinary System, located near the spine at the small of the back. The kidneys extract water and urea, mineral salts, toxins, and other waste products from the blood with filtering units called nephrons. From the nephrons the collected waste (Urine) is sent to the bladder for excretion. One Kidney must function properly for life to be maintained. Kidney diseases include nephritis, inflammation of the kidneys caused by infection of degenerative changes in the renal capillaries, and nephrosis, a condition resulting from lesions in the renal tubules characterized by general edema and by protein in the urine.

Kieselguhr. A porous form of silica, SiO_2, formed from skeletons of microscopic plants. It is used as an absorbent.

Killed spirits. A solution of zinc chloride, $ZnCl_2$, made by adding pieces of zinc to spirits of salt (hydrochloric acid). It is used as a soldering flux.

Kiln. Furnace for firing pottery and enamels; for making brick, charcoal, lime and cement; for roasting ores; and for drying various substances. A kiln may be fired intermittenly or continuously. Rotary kiln, much used in continuous processes, e.g., cement manufacturing, consist of long tubes, lying almost horizontal, that are rotated slowly as heat is applied to the material being treated inside the tubes.

Kilo-. Prefix indicating one thousand (10^3).

Kilocycle. A unit of frequency equal to 1000 cycles per second (1000 hertz).

Kilogram (kilogramme). Symbol : kg. The basic SI unit of mass, equal to the mass of the *international prototype kilogram*—a cylindrical piece of platinum-iridium alloy kept in the International Bureau of Weights and Measures at Sevres near Paris. It is equivalent to 2.204 62 pounds.

Kilohertz. Symbol : kHz. A unit of frequency equal to 1000 hertz.

Kilometre. A unit of distance equal to 1000 metres. It is equivalent to 0.6214 mile.

Kilowatt. Symbol : kW. A unit of power equal to 1000 watts.

Kilowatt-hour (Board of Trade unit). Symbol : kWh. A unit of energy equal to a power of one kilowatt available for one hour. It is used for the sale of electrical energy.

Kinematics. The branch of mechanics concerned with motion without reference to the forces or mass involved in the motion. *Compare* dynamics.

Kinematic viscosity. *See* viscosity.

Kinetic energy. Symbol : E_K. The energy that a system has by virtue of its motion, determined by the work necessary to bring it to rest. An object of mass *m* and velocity *v* has a kinetic energy equal to mv^2. A body rotating about an axis with moment of inertia I and angular velocity *w* has a kinetic energy equal to to $Iw^2/2$.

Kinetic-molecular theory of gases. Physical theory that explains the behavior of gases by assuming that any gas is composed of a very large number of very tiny particles, called molecules, that are very far apart compared to their sizes. The molecules are assumed to exert no force on one another, except during rare, perfectly elastic collisions. A gas corresponding to these assumptions is called an ideal gas. The analysis of the behavior of an ideal gas according to the laws of mechanics leads to the GAS LAWS. The theory also shows that the absolute TEMPERATURE is directly proportional to the average kinetic energy of the molecules. Pressure is seen to be the result of large numbers of collisions between molecules and the walls of the container in which the gas is held. See Thermodynamics.

Kinetics. The study of the rates of chemical reactions, used to study reaction mechanisms.

Kinetic theory. Any theory for describing the physical properties of a system with reference to the motion of its constituent atoms or molecules. The most successful is the *kinetic theory of gases,* which is based on the principle that gases are composed of many elastic particles of negligible size in constant random motion. The particles collide with each other and with the walls of the containing vessel, exerting a pressure of $mvC^2/3$, where m is the particle mass, C the root-mean-square velocity, and v the number of particles per unit volume.

Kinship. Relationship between persons by blood or Marriage; also, a system of rules based on such relationships and governing marriage, descent, inheritance, and, sometimes, residence. All societies recognize consanguinity (relationship by blood or by descent from a common ancestor) and affinity (relationship to a spouse's cosanguineous relatives). Marriage between persons in lineal consaguinity (in direct line of descent, e.g., father and daughter) or between brother and sister is considered incest and is almost universally void under church, civil, and common law, while marriage between persons of collateral consanguinity (sharing a common ancestor but not in direct line of descent, e.g., first cousins) may be allowed. Every individual is classified by kinship lines, or descent. Social organizations often dominated by descent. Social organization is often dominated by descent groups based on links through females (matrilineal), males (patrilineal), or both (bilateral). In some societies the concept of kinship extends beyond the family to groups or clans in which consanguinity is hypothetical or even mythological, but which are governed by all the laws of descent.

Kipp's apparatus. A laboratory apparatus for generating gas by chemical reaction of a liquid and a solid when the tap is closed the pressure of gas drives the liquid into the top bulb, thus terminating the reaction. [After Petrus Kipp (1808-64).]

Kirchhoff, Gustav Robert (kirkh'hof), 1824-87. German physicist. He was professor of physics at the universities of Breslau, Heidelberg, and Berlin. He and Robert Bunsen, working with the spectroscope, discovered the elements Cesium and Rubidium. Kirchhoff explained the Fraunhofer lines in the solar spectrum and formulated Kirchhoff's Laws describing current and voltage in an electric circuit.

Kirchhoff's laws. Pair of laws stating general restrictions on the current and voltage (see Potential, Electric) in an Electric Circuit or network. The first states that at any given instant the sum of the voltages around any closed path, or loop, in the network is zero. The second states that at any junction of paths in a network the sum of the currents arriving at any instant is equal to the sum of the currents flowing away.

Kirchhoff's law. The principle that the emissivity of a body is equal to its absorptance at the same temperature. [After Robert Kirchhoff (1824-87), German physicist.]

Kirchhoff's laws of (electricity). Two laws applied to the flow of electric currents in a network.

1. The algebraic sum of the currents at any point in the network is zero—a current flowing towards the point is taken as having opposite sign to one flowing away from the point.

2. The algebraic sum of the e.m.f.s in any closed loop of the network is equal to the algebraic sum of the products of the currents and the resistances through which they flow : i.e., $E_1 + E_2 + ... = I_1R_1 + I_2R_2 + ...$

Kitt Peak National Observatory. Astronomical observatory (founded 1960) located at an altitude of 7,000 ft (2,130 km) on the Papago Indian reservation SW of Tucson, Ariz., it is administered by the Association of Universities for Research in Astronomy under contract with the U.S. National Science Foundation. Principal instruments include the Mayall 158-in. (4.01-m) reflector and the 60-in. (1.52-m) Robert McMath Solar Telescope, the largest of this type. The observatory's ten telescopes are supplemented at Kitt Peak by four other telescopes operated by other institutions.

Kjeldahl's method. A technique for determining the amount of nitrogen in organic compounds by decomposing the sample with concentrated sulphuric acid to convert the nitrogen into ammonium sulphate. The amount of ammonium sulphate produced is measured by adding an excess of alkali, distilling off ammonia into a standard acid solution, and titrating the excess acid. [After Johan Kjeldahl (1849-1900), Danish chemist.]

Klystron. An electron tube for amlifying or generating radiofrequency electromagnetic radiation. A beam of electrons is passed through two consecutive resonant cavities. High-frequency waves generated in the first cavity cause the electrons to "bunch" into a series of pulses. These generate more high-frequency waves, of higher amplitude, in the second cavity, the energy being supplied by the electron beam. The process of bunching the electrons is called *velocity modulation.*

Klein, (Christian) Felix, 1849-1925. German mathematician, noted for his work in geometry and on the theory of functions. His "Erlangen Programm" (1872) for unifying the diverse forms of geometry through the study of equivalence in transformation groups was influential for over 50 years. Klein was a prolific writer and lecturer on the theory, history, and teaching of mathematics.

Knitting. Construction of fabric by interlocking loops of thread or yarn using long needles. Hand knitting begins with the casting on a row of loops, into which successive rows of stitches are worked, moving alternately from one needle to the other. All stitches and patterns derive from theknit and purl stitches. With fine needles and thread a form of Lace can be made. Knitting was unknown in Europe before the 15th cent. The first knitting machine was invented in England in 1589. Modern automatic machines are of two types, warp and weft. In warp knitting the fabric is constructed of vertical chains made up of separate threads and connected by zigzag stitches. Weft knitting (which includes hand knitting) uses one continuous thread to produce a horizontally constructed fabric. Warp-knit fabrics run-resistant but less elastic than weft-knit.

Knot. A unit of velocity equal to one nautical mile per hour (1.15 statute miles per hour.) It is used for expressing the speeds of ships and aircraft.

Knudsen flow. *See* Molecular Flow. [After M. Knudsen (1871-1949), Danish physicist.]

Koala. Arboreal, nocturnal Marsupial (*Phascolarctos cinereus*) native to Australia. Bearlike in appearance, it has thick, gray fur, a tailless body 2 to $2^1/_2$ ft (60 to 75 cm) long, a black nose, and large, furry ears. Its diet consists of only one species of eucalyptus (see Myrtle) at a particular stage of maturation. The Koala has been hunted for food and fur and is now endangered.

Koch, Robert, 1843-1910. German bacteriologist. He devised a method of staining bacteria with aniline dyes and developed bacteriological culture techniques still used. He established the bacterial cause of many infectious diseases, including anthrax (1876), tuberculosis (1882), conjunctivitis (1883), and Asiatic cholera (1884), and studied sleeping sickness, malaria, bubonic plague, rinderpest, and other diseases. For developing tuberculin as a test of tuberculosis he received the 1905 Nobel Prize in Physiology or medicine.

Kornberg, Arthur, 1918-. American biochemist. For his discovery of the mechanisms in the biological synthesis of DNA, he shared with Severo Ochoa the 1959 Nobel Prize in physiology or medicine. He has done research at the National Institutes of Health (1942-52), Washington Univ. (1953-59), and Stanford Univ. (1959-).

Korolev, Sergei Pavlovich, 1907-66. Soviet rocket designer. Korolev's role as the chief designer of the Soviet space program was not revealed until after his death. After he successfully launched two intercontinental ballistic missiles in Aug. 1957, Soviet premier Nikita Khrushchev allowed him to attempt the orbiting of a test satellite; as a result,

Sputnik 1, the first artificial earth satellite, was orbited successfully on Oct. 4, 1957. Korolev also directed the launching of the USSR's first unmanned space probes and manned space missions.

Kosher. [Heb., = proper], term applied to food that is in accordance with Jewish dietary laws. Animals that chew their cud and have cloven hooves (e.g., cows and sheep) are kosher; rules governing their slaughter and preparation also apply to fowl. Proper preparation of kosher meats includes complete drainage of blood. Kosher fish are those with scales and fins. Milk products may not be cooked or eaten with, or immediately after, meat or poultry. The origins of and motivations for Jewish dietary laws and customs have been variously explained as hygenic, aesthetic, folkloric, ethical, and psychological.

Kovalevsky, Sonya or **Sophie,** 1850-91. Russian mathematician. A student of Karl Weierstrass, she received (1874) a ph.D. from the Univ. of Gottingen for a remarkable thesis on partial differential equations. She won (1888) a French Academy of Sciences prize for a memoir on the rotation of a solid body about a fixed point.

Krebs, Sir Hans Adolf, 1900-1981. English biochemist; b. Germany; emigrated to England, 1933. For his studies of intermediary metabolism he shared with Fritz Lipmann the 1953 Nobel Prize in physiology or medicine. These studies included the elucidation of the cycle of chemical reactions called the Citric Acid Cycle, or Krebs cycle, the major source of energy in living organisms.

Kroll process. A method of producing metallic elements by reduction of their chloride with magnesium.

Krupp (kroop). Family of German armament manufacturers who flourished in the 19th and 20th cent. *Friedrich Krupp,* 1786-1826, built a small steel plant at Essen c. 1810. His son, *Alfred Krupp,* 1812-87, the "cannon king," specialized in armaments and acquired mines all over Germany. Under Alfred's son, *Friedrich Alfred Krupp*, 1854-1902, the family vastly extended its operations. The husband of his daughter, Bertha Krupp (after whom the Big Bertha guns were named), took the name *Gustav Krupp von Bohlen und Halbach*, 1870-1950. He took over that what was then a public company and made it the center of Nazi rearmament in the 1930s. In 1943, Hilter converted the company back into a family holding, and *Alfred Krupp von Bohlen und Halbac*, 1907-67, son of Gustav and Bertha, ran the company until the end of World War II. He was convicted (1948) as a war criminal but was released from prison (1951) and allowed to resume control of the firm. After his death, Krupp became a public corporation and the Krupp family ceased (1968) to control it.

Krypton. Symbol : Kr. A colourless odourless, tasteless, nonflammable gaseous element present in the atmosphere (0.5 parts per million) from which it is extracted as a by-product in the liquefaction of air. It is used in fluorescent lamps and lasers. Krypton is one of the inert gases. A.N. 26; A.W. 83.80 ; m.pt. - 156.6°C; b.pt. - 152.3°C.

Krypton (Kr). Gaseous element, discovered by William Ramsay and M.W. Travers in 1898. It is a rare Inert Gas used to fill electric lamp bulbs and various electronic devices, and to detect heart defects. The definition of a meter is based on the emission spectrum of the krypton-86 isotope.

Kubler-Ross, Elisabeth, 1926-. Swiss-American physician; M.D., Univ of Zurich, 1957. After settling in the U.S. in 1958, she taught psychiatry and specialized in treating the terminally ill. Her best-selling book *On Death and Dying* (1969), describing five psychological stages experienced by the dying (denial, anger, "bargaining for time," grieving, and acceptance), has greatly influenced the humane treatment and counseling of terminally ill patients and the rise of Hospice care for dying persons.

Kuhn, Richard, 1900-1967. Austrian chemist. Awarded the 1938 Nobel Prize in chemistry for his work on the carotinoids and on vitamins, he was unable, because of the Nazis, to accept the award until after World War II. He isolated vitamins B_2 (riboflavin) and B_6.

Kumquat. Evergreen shrub (genus *Fortunella*) of the Rue family, cultivated for its small, orange-yellow Citrus Fruits, eaten fresh or in preserves. Kumquats, with their sweet-scented white flowers, are also grown as ornamentals in warm areas of the U.S.

Kwashiorkor. Protein deficiency disorder of children, prevalent in overpopulated parts of the world where the diet consists mainly of starchy vegetables, particularly Africa, Central and South America, and S. Asia. Such a diet is deficient in certain amino acids, which make up Proteins vital for growth. Depending on the extent, onset, and duration of the deficiency, manifestation include skin changes, edema, severely bloated abdomen, diarrhea, and generally retarded development.